Fixed Points

Mathematical World • Volume 2

Fixed Points

Yu. A. Shashkin
Translated from the Russian by
Viktor Minachin

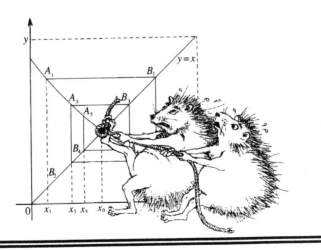

American Mathematical Society
Mathematical Association of America

Ю. А. ШАШКИН

НЕПОДВИЖНЫЕ ТОЧКИ

«НАУКА», МОСКВА, 1989

Translated from the Russian by Viktor Minachin

1991 *Mathematics Subject Classification*. Primary 01-01, 54-01, 54H25; Secondary 54C15.

Library of Congress Cataloging-in-Publication Data

Shashkin, ÎU. A. (ÎUriĭ Alekseevich)

[Nepodvizhnye tochki. English]

Fixed points/Yu. A. Shashkin; translated from the Russian by Viktor Minachin.

    p. cm.—(Mathematical world, ISSN 1055-9426; v. 2)

Includes bibliographical references.

ISBN 0-8218-9000-X (alk. paper)

1. Fixed point theory. I. Title. II. Series.

QA329.9.S5313 1991   91-28994

515$'$.7248—dc20   CIP

Table of Contents

Preface

Applying mathematics means, in many cases, solving equations. If that is the case, then the important thing to know is whether a particular equation has a solution or not. The presence of solutions is guaranteed by so-called existence theorems. Let f be a function of the real variable x, continuous in the closed interval $[a, b]$ and assuming values of different signs at its endpoints. Then the equation

$$(0.1) \qquad f(x) = 0$$

has at least one solution inside the interval.

Existence theorems are often expressed in the form of "fixed point" principles. For example, let us view equation (0.1) in the following way. Write (0.1) in the form $\lambda f(x) + x = x$, where λ is a positive parameter. Denoting $\lambda f(x) + x$ by $F(x)$ we get the equation

$$(0.2) \qquad F(x) = x.$$

Choose the value of λ in such a way that all the values of F lie inside the interval $[a, b]$. Equation (0.2) can now be looked upon as follows. The function F maps the point (the real number) x from the interval $[a, b]$ into the point $F(x) = y$ of the same interval, which, in general, does not coincide with x. In other words, the mapping F takes the point x into the point y. However, if a point x_0 is a solution of (0.2), then it stays where it was; that is, it is a *fixed* point. The same point is evidently a solution of equation (0.1) as well.

Therefore, in geometrical terms, a theorem ensuring the existence of a solution for equation (0.2) is formulated as the following *fixed point principle*: if F is a continuous function mapping a closed interval into itself, then the function has at least one fixed point. Because F is an arbitrary continuous function, this property is actually that of the closed interval itself; it does not depend on the choice of a particular mapping and is called the fixed point property.

The problem of solving the system of two equations

$$f(x, y) = a, \qquad g(x, y) = b$$

for unknown x and y can be reduced to the problem of whether a mapping of a square or, say, a disk into itself has a fixed point.

Fixed point theorems have numerous applications in mathematics. Most of the theorems ensuring the existence of solutions for differential, integral, operator, or other equations can be reduced to fixed point theorems. They are also used in new areas of mathematical applications, e.g., in mathematical economics, game theory, etc.

The subject of this book is essentially one single problem: whether a closed interval, square, disc, or sphere has the fixed point property.

The theory of fixed points belongs to topology (see [3] and [5]), a part of mathematics created at the end of the nineteenth century, and makes extensive use of such topological notions as continuity, compactness, homotopy, and the degree of a mapping.

Another aim of the book is to show how combinatorial considerations related to decomposition (triangulation) of figures into distinct parts called faces (simplexes) adjoining each other in a regular fashion are used in this theory.

Three names should be mentioned here.

The first is the famous French mathematician H. Poincaré (1854–1912), the founder of the fixed point approach, who had deep insight into its future importance for problems of mathematical analysis and celestial mechanics, and took an active role in its development. Poincaré was the first to apply the combinatorial approach to topology, using triangulations of geometrical figures into simplexes.

The second is the Dutch mathematician L.E.Y. Brouwer (1881–1966). He introduced the topological notions used in this book, among them those of homotopy and the degree of a mapping. He also proved the fixed point theorems for a square, a sphere, and their n-dimensional counterparts.

The third is the German mathematician E. Sperner (1906–1980), who back in 1928 proved the combinatorial geometric lemma on the decomposition of a triangle (as well as of any n-dimensional simplex in general), which plays an important role in the theory of fixed points.

The author would like to express his deep gratitude to V. G. Boltyansky, whose suggestions led to substantial improvements of the first version of the manuscript. He is also indebted to E. G. Pytkeyev for useful discussions during the preparation of this book.

1

Continuous Mappings of a Closed Interval and a Square

Consider a closed interval $I = [0, 1]$ of the real line $\mathbf{R}$. Let f be a function mapping I onto itself. This means that a rule is specified associating to each point x in I a unique point $f(x)$ of the same interval. We say that $f(x)$ is the *image* of the point x or that the point x goes into the point $f(x)$. Along with the term "function" we will also use the term "mapping"; for us they are synonyms. The only condition the mapping f should satisfy is that it be *continuous*. The meaning of continuity is intuitively clear: the graph of the mapping is a continuous curve and can be drawn without lifting a pen off the paper. Another way to describe a continuous mapping is to say that it takes close points into close ones. For the moment we will be satisfied with the general idea and postpone a precise definition until Chapter 5.

Consider a square Q lying on the coordinate plane $\mathbf{R}^2$ so that the coordinates of its points satisfy the following inequalities: $0 \le x \le 1$, $0 \le y \le 1$ (see Figure 1). Let f be a continuous mapping of the square Q into itself. This means that each point p of Q goes into a uniquely defined point $f(p)$ of the same square, and the images of close points are also close to each other. To define such a mapping one must specify two real continuous functions defined on the square Q and satisfying the inequalities

$$(1.1) \qquad 0 \le g(x, y) \le 1, \qquad 0 \le h(x, y) \le 1.$$

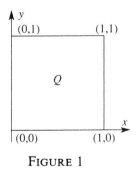

FIGURE 1

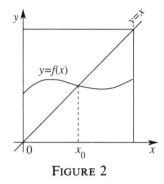

FIGURE 2

If (x_0, y_0) is a point of the square, then its image under the mapping f is the point $(g(x_0, y_0), h(x_0, y_0))$ that, according to inequalities (1.1), also lies inside the square.

Let us make the difference between the notions of a *mapping* and a *function* more precise. The first term will be used in a wider sense, as a mapping of any set into any other (or the same) set. The term "function" will always mean a mapping of any set into the real line.

The Brouwer fixed point theorem states that *any continuous mapping f of a closed interval into itself has at least one fixed point, i.e., a point x_0 such that $f(x_0) = x_0$; any continuous mapping of a square into itself also has a fixed point.*

Note that the only requirement imposed on the function f is that it should be continuous—the statement is of a very general nature indeed.

Here is an intuitively obvious proof of this theorem for a closed interval (see Figure 2). The graph of the function f is a continuous curve joining a point on the left side of the square with a point on the square's right side. It must, therefore, intersect the diagonal of the square. But the coordinates $(x_0, f(x_0))$ of any point of the graph lying on the diagonal satisfy the condition $f(x_0) = x_0$.

For a similar intuitively clear proof in the case of a square, we have to visualize the graph of a mapping f of a square into itself. Each point of a square is described by two coordinates. The same is true its image. Therefore, the pair consisting of a point and its image is described by four coordinates. In other words, the graph of the mapping f is a curved surface in a 4-dimensional space. It is not an easy task to visualize such a curve, so we will use another approach and consider two drawings instead of one.

Now here is the proof. Figure 3a shows the original square $ABCD$. The rectangle $ADGF$ bounds a region of the plane $z = x$ lying above the square $ABCD$. (In other words, $ADGF$ is the graph of the function of two variables equal to x.) Here $IJKL$ is a continuous surface $z = g(x, y)$ (the graph of the function equal to $g(x, y)$); MN is the intersection curve of the plane $z = x$ and the surface $z = g(x, y)$; the curve PQ is the projection of MN on the plane $z = 0$. The curve PQ joins the opposite sides AB and

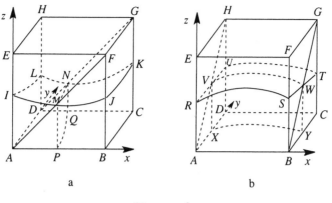

FIGURE 3

CD of the square, and coordinates of any point on CD satisfy the equation $g(x, y) = x$.

Similarly, in Figure 3b the rectangle $ABGH$ is the graph of the function $z = y$; the surface $RSTU$ is the graph of the function $z = h(x, y)$; VW is the intersection curve of both surfaces, and XY is the projection of this curve on the plane $z = 0$. The continuous curve XY joins the opposite sides BC and AD of the square, and the coordinates of each of its points satisfy the equation $h(x, y) = y$. It is now clear that the curves PQ and XY must intersect at least at one point (x_0, y_0) and that this point is a fixed point because it satisfies the equations $g(x_0, y_0) = x_0$ and $h(x_0, y_0) = y_0$.

For a rigorous proof of the Brouwer theorem, we need new notions and facts. First, there are combinatorial lemmas on the decomposition of a closed interval and square (or a triangle) into a finite number of smaller parts. Second, we have to make the notion of continuity more precise. Third, we need the notion of compactness.

Problems. 1. The functions

$$f_1(x) = 2x, \quad f_2(x) = \sin x, \quad f_3(x) = x^2 - \frac{x}{2} + \frac{1}{2}, \quad f_4(x) = \frac{1}{1 + x^2}.$$

are defined on the real line **R**. Which function maps the closed interval $[0, 1]$ into itself?

2. For each of the following sets give an example of a continuous mapping of the set into itself that has no fixed points: (a) the real line, (b) the interval closed at one end $(0, 1]$, (c) the pair of closed intervals: $[-2, -1]$ and $[1, 2]$.

3. Give an example of a mapping of the closed interval $[0, 1]$ into itself that is not continuous and has no fixed points.

4. Let f be a continuous function. The function f^2 is defined by the expression $f^2(x) = f[f(x)]$, i.e., in order to find $f^2(x)$ one has to evaluate the function f at the point $f(x)$. A continuous mapping f of the closed

interval [0, 1] into itself is said to be an *involution* if any point of the interval
is a fixed point for f^2. In other words, the function f equals its inverse.
   Prove that the following functions are involutions

$$f_1(x) = x, \quad f_2(x) = 1 - x, \quad f_3 = \frac{1-x}{1+x},$$

$$f_4(x) = \sqrt{1 - x^2}, \quad f_5(x) = 1 - \sqrt{1 - (x-1)^2}.$$

Here $\sqrt{}$ means the positive square root.
   The involution $f(x) = x$ is said to be trivial.
   Prove that any nontrivial involution has exactly one fixed point.

2

First Combinatorial Lemma

This is the name of Lemma 1. It is also known as Sperner's Lemma for a closed interval.

LEMMA 1. *Suppose that a finite number of points subdivides a closed interval into smaller intervals. The left endpoint of the original interval is labeled by* 0, *the right one by* 1, *and each of the partitioning points inside the interval is also labeled by either* 0 *or* 1. *Then there is an interval of the subdivision whose endpoints are labeled by different numbers. Moreover, the number of such intervals is odd.*

PROOF. Let us call a small interval "acceptable" if its endpoints are labeled differently. There are only two possibilities. Either all the interior points are labeled by 0 or at least one of them is labeled by 1. In the first case there is exactly one "acceptable" interval, the extreme right one (see Figure 4a). In the second case consider the extreme left of all the points labeled by 1. Evidently, the interval for which this point is the right endpoint is an "acceptable" one (see Figure 4b).

Now let us show that the number of "acceptable" intervals is necessarily odd. Start moving along the original interval from left to right, counting "acceptable" intervals on our way: one, two, etc. Each odd "acceptable" interval has its left endpoint labeled by 0 and its right endpoint labeled by 1, while the endpoints of even "acceptable" intervals are labeled the other way around. The right endpoint of the last "acceptable" interval is labeled 1, hence the number of "acceptable" intervals is odd. The lemma is proved.

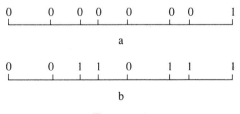

FIGURE 4

The statement of the lemma can be made even more precise. Let us say that an "acceptable" interval is of type $(0, 1)$ if its left endpoint is labeled by 0 and its right endpoint is labeled by 1. Otherwise the interval is of type $(1, 0)$. Then the number of intervals of type $(0, 1)$ is one more than the number of those of type $(1, 0)$. The reader can easily verify this statement.

3

Second Combinatorial Lemma, or Walks through the Rooms in a House

We will formulate this lemma as a statement about the rooms and the doors in a house (see Figure 5). Suppose that the number of doors in each room equals 0, 1, or 2. A room with a single door is called a *dead end,* a room with two doors is a *communicating room.* A door can be either an *outside* one, if it leads out of the house, or an *inside* one, if it connects two neighboring rooms. It is also quite natural to assume that no room has more than one outside door, and no two neighboring rooms have more than one common door.

LEMMA 2. *Suppose that any room of a house has 0, 1, or 2 doors. Then the number of dead ends and the number of outside doors are of the same parity.*

The last sentence means that both of them are either even or odd simultaneously.

PROOF. For the proof let us describe the walks through the rooms of the house. Each walk will be made according to the following rules. First, any door can be passed only once. Second, a walk starts either by entering the house from the street through an outside door or from a dead end. The walk continues through communicating rooms and terminates in one of two cases. Either we get outside or find ourselves in a dead end (see Figure 5). The assumptions on the number of doors in each room determine the walk uniquely. After entering a two-door room, one can exit only one way. After

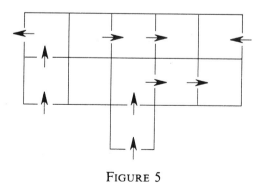

FIGURE 5

the walk is finished, we start another one, and continue until there are no more outside doors or dead ends to start from.

The resulting paths are of one of three types:

(1) from an outside door into a dead end (or the other way round, which for our purposes makes no difference);

(2) from an outside door to another outside door;

(3) from a dead end into a dead end.

Denote by m, n, and p the numbers of paths for the first, second, and third type, respectively.

Because each path of type (1) corresponds to one outside door, and each path of type (2) corresponds to two outside doors, the total number of outside doors is $m + 2n$. Similarly, the number of dead ends is $m + 2p$. The numbers $m + 2n$ and $m + 2p$ are of the same parity, which proves the lemma.

Note that the shapes of the room are irrelevant. They can, for example, be triangular, and this is what we shall assume next.

4

Sperner's Lemma

Consider an arbitrary triangle subdivided into smaller triangles. We will always assume that the subdivision satisfies the following condition: any pair of smaller triangles either have no common points, or have a common vertex, or have a common side. Such a subdivision is called a *triangulation,* smaller triangles are said to be the *faces* of the triangulation, the sides of smaller triangles are its *edges,* and their vertices are the *vertices* of the triangulation.

For example, the subdivision shown in Figure 6a is a triangulation, while that in Figure 6b is not.

LEMMA 3 (Sperner). *Consider a triangulation of a triangle* T. *The vertices of this triangle are labeled by* 1, 2, *and* 3. *The vertices of the triangulation are labeled by the same numbers in such a way that the following boundary condition is satisfied: if a vertex lies on a side of the triangle* T, *it is labeled by one of the two numbers labeling the endpoints of this side. Then there is at least one face in the triangulation with the vertices labeled by different numbers, i.e.,* 1, 2, *and* 3. *Moreover, the number of such faces is odd.*

PROOF. An illustration to Lemma 3 is given in Figure 7 on p. 10. It is proved by reduction to Lemma 2. Let us call triangle T a house and each face of the triangulation a room. An edge of the triangulation is called a door provided its endpoints are labeled by 1 and 2. We shall say that such edges are of type (1, 2). (Note that, in contrast to Lemma 1, we make no difference

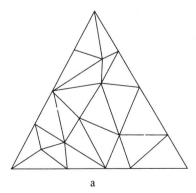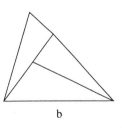

a   b

FIGURE 6

9

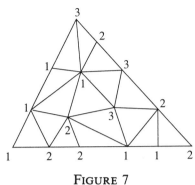

FIGURE 7

between edges of type (1, 2) and (2, 1).) Which faces are now the *dead ends*? Consider all possible assignments of labels 1, 2, and 3 to the vertices of a face. They can be of ten different types: (1, 1, 1), (1, 1, 2), (1, 1, 3), (1, 2, 2), (1, 2, 3), (1, 3, 3), (2, 2, 2), (2, 2, 3), (2, 3, 3), (3, 3, 3). Evidently, only the faces of type (1, 2, 3) have exactly one edge of the type (1, 2), so it is quite natural to call them dead ends. Similarly, the faces of types (1, 1, 2) and (1, 2, 2) should be called *communicating rooms,* because they contain exactly two edges of the type (1, 2). We, therefore, obtain the following "glossary" for translating conditions of Lemma 3 into the language of Lemma 2:

triangle *T*—house

face of triangulation—room

edge of triangulation of type (1, 2)—door

boundary (resp., inside) edge of triangle *T*—outside (resp., internal) door

face of type (1, 2, 3)—dead end

face of type (1, 1, 2), or (1, 2, 2)—communicating room

Whatever the choice of labels for the vertices of a triangulation is, the list shows that a face may have one or two edges of type (1, 2), or none at all. Hence, the conditions of Lemma 2 are satisfied. Therefore, the number of dead ends and outside doors is either both even or both odd. Translating this back into the language of Lemma 3, we conclude that the number of faces of type (1, 2, 3) and the number of boundary edges of type (1, 2) are of the same parity. It remains to show that the second number is odd. But the boundary edges of type (1, 2) can lie only on the side of the triangle *T* labeled by 1 and 2. Because there are no edges of this type on the other sides of the triangle, the required conclusion follows from Lemma 1. Lemma 3 is proved.

The process used in the proof will be called the *walking process.* Below we shall use it by "walking" the faces of a triangulation itself, without necessarily calling them rooms. The crucial point will always be what is meant by a "door" edge, and by a "dead end" face.

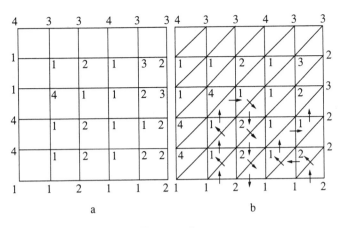

FIGURE 8

Let us use this approach to prove Lemma 4, which makes the same statement as Lemma 3, but for a square. Let a square Q be divided into smaller squares by lines parallel to its sides. For this subdivision we will use the same names as before, i.e., faces, edges, and vertices of a subdivision, where, for example, faces are smaller squares of the subdivision.

LEMMA 4. *Let a square Q be subdivided into smaller squares by lines parallel to its sides. The vertices of the square are labeled by numbers 1, 2, 3, and 4. The vertices of the subdivision are labeled by the same numbers in such a way that the following boundary condition is satisfied: if a vertex of the subdivision lies on the side of the square Q, then it is labeled by one of the two numbers labeling the endpoints of this side (see Figure 8a). Then there is at least one face labeled by at least three different numbers.*

PROOF. To prove the theorem let us subdivide each square face into two triangular faces (see Figure 8b). The result is a triangulation of the square, all vertices of which are labeled. Let us call an edge with the labels (1, 2) a *door*. A *dead end* is a triangular face labeled by three different numbers such that at least two of them are 1 and 2. Thus, dead ends are triangles of type (1, 2, 3) or (1, 2, 4).

By the hypothesis of Lemma 4, all the outside doors lie on side (1, 2) of the square, and by Lemma 1 their number is odd.

Any walk through the triangular faces of the square starts at an outside door and ends in either of the two cases: leaving the square through another outside door or entering a dead end (see Figure 8b). Since the total number of outside doors is odd, at least one of the walks must end in a dead end. This means that there is a triangular face labeled either by (1, 2, 3), or (1, 2, 4), and, consequently, there is a square face labeled by at least three different numbers.

Problems. 5. Prove the first combinatorial lemma by a walking process. Consider a closed interval as a house in Lineland. What are its rooms, doors, and dead ends?

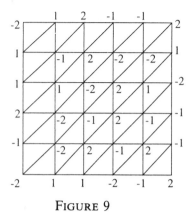

FIGURE 9

6. A triangulation of square Q is given, as in Figure 8b, or Figure 9. Each of the vertices of the triangulation is labeled by one of the following numbers: $1, -1, 2, -2$. The boundary condition is that the vertices lying opposite each other on the boundary (i.e., if the line passing through them goes through the center of Q) are labeled by opposite numbers (see Figure 9). Prove that there is an edge of the triangulation whose endpoints are labeled by opposite numbers, i.e., either an edge of type $(1, -1)$ or an edge of type $(2, -2)$.

7. The squares of a chess board (either a usual 8 by 8 board, or any other rectangular board of size m by n) are divided between a king and a rook in such a way that both pieces can move (according to the usual rules) just over its own squares (see Figure 10). Suppose there is no square on the extreme left-hand side of the chessboard from which the king can get to the extreme right of the board. Prove that then the rook can always find its way from the bottom row to the top row.

Is the statement true if the king is replaced by a queen? And if both pieces are rooks?

8. (The "Hex" game.) The Hex game is played on a rhombus-shaped board made up of regular hexagons (see Figure 11). The board usually consists of 11 by 11 hexagonal fields, but for our purposes it may be of any size. A pair of opposite sides of the board belongs to the player who has the white pieces,

r	k	k	k	k	k	k	r
k	r	r	k	r	r	r	r
r	k	r	k	r	r	k	k
k	k	k	k	k	r	k	r
r	r	r	r	r	r	k	r
k	r	k	k	k	k	r	k
k	r	r	r	k	k	r	k
k	k	k	r	r	k	k	r

FIGURE 10

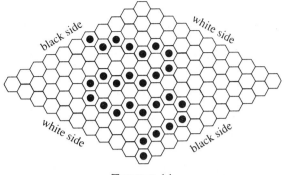

FIGURE 11

and the sides are therefore called "white." The other two sides of the board belong to the player who has black pieces, and are called "black." The four corner fields belong to the two sides they lie in. Each player in turn places a piece of his or her color on any free field of the board. The goal of each player is to join his or her sides of the board with an unbroken chain of his or her pieces. For example, in Figure 11 the blacks have won. Prove that any hex game must end with a winner, i.e., there can be no draws.

9. A square Q is divided into smaller squares (faces) by lines parallel to its sides. The boundary of the square Q is *oriented*. That is, a direction is chosen on it such that if you move in that direction along the boundary, the internal points of the square Q always stay on your left (or on your right). For each of the other edges of the subdivision a direction is specified such that for each vertex inside Q there are exactly two edges coming to the vertex and two edges going from it (see Figure 12).

Prove that there is at least one oriented face in the subdivision (in Figure 12 there is exactly one such face, namely in the extreme lower right).

10. A *polygon* is a flat figure made up of a finite number of triangles (faces) in such a manner that (a) any two faces have no points in common or have a common vertex or a common side (edge); (b) any vertex can be reached from any other vertex by moving along the edges; (c) the figure has no holes inside. Thus, we assume that a polygon is defined by its triangulation.

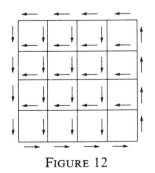

FIGURE 12

The *multiplicity* of a vertex is the number of edges for which it is an endpoint. Prove that the number of vertices of odd multiplicity in a triangulation of a polygon is even.

11. In Sperner's lemma we are looking for the faces of a triangulation labeled by three different numbers, for example, 0, 1, and 2. Consider a triangulation such that (a) each of its faces is labeled by three different numbers 0, 1, and 2. Prove that then (b) all faces of the triangulation can be marked by two colors (black and white) so that no two adjacent faces (i.e., faces having a common edge) are marked by the same color. Prove that, conversely, (b) implies (a).

12. A triangle is triangulated in the following manner: each of its sides is divided into m equal parts (m is an arbitrary natural number), and the lines parallel to the sides of the triangle are drawn through the subdividing points (see Figure 13).

Each vertex of the triangulation is colored either red or blue. Prove that the number of edges of the triangulation having endpoints of different colors is even.

Does the statement hold for any triangulation of the triangle? If not, what is the class of triangulations for which it is valid?

FIGURE 13

5

Continuous Mappings, Homeomorphisms, and the Fixed Point Property

As we have already said, a mapping is continuous if it takes points close to each other into points that are also close to each other. Now we have to make this definition more precise. Whether points are close to each other or not depends on the distance between them. A point y is said to be close (or sufficiently close) to the point x if the distance between them is less than a fixed number $r > 0$. The set of such points is said to form the r-neighborhood of the point x. In geometrical terms, an r-neighborhood on a plane is a disc without boundary of radius r with center at x. In three-dimensional space a neighborhood is a ball without its boundary surface. If X is a plane set (line, circle, square, etc.), and x is one of its points, then the r-neighborhood of the point x in the set X is, by definition, the set of all points in X whose distance from x is less than r. Therefore, such a neighborhood is the intersection of the set X and the disc of radius r (see Figure 14).

Now let X and Y be two sets (on a plane) and f a mapping from X to Y. This is denoted as follows: $f : X \to Y$. If x is a point of X and $y = f(x)$, then y, as we already know, is said to be the *image* of the point x, while the point x is called the *preimage* of the point y.

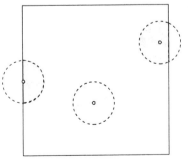

FIGURE 14

The mapping $f : X \to Y$ is said to be continuous at a point x if for any $\varepsilon > 0$ there exists a number $\delta > 0$ such that the image of each point in X from the δ-neighborhood of the point x lies in the ε-neighborhood of the point y.

Thus, the continuity of a mapping at a point x means that for any arbitrary neighborhood of the image point y there is a neighborhood of the point x, corresponding to the first neighborhood and depending on it. Here the term "corresponding" means that all points of the second neighborhood are mapped into the first one.

If a mapping $f : X \to Y$ is continuous at each point of the set X then it is said to be *continuous on* X or simply continuous.

Example 1. The projection of a square on its side AB is the mapping taking each point x of the square into the endpoint $f(x)$ of the perpendicular dropped from x on AB (see Figure 15). For any ε-neighborhood of the point $f(x)$ the δ-neighborhood of the point x of the same radius projects into the first neighborhood. Hence, for any $\varepsilon > 0$ there exists a corresponding $\delta > 0$, in this case equal to ε. Therefore the projection map is continuous at each point of the square.

Example 2. Let K_1 and K_2 be two concentric discs with a common center at the point O and radii r and $R > r$, respectively. Define a mapping $f : K_1 \to K_2$ as follows. The point O goes into itself, while each radial segment OA of the disc K_1 is mapped linearly onto the radial segment OB of the circle K_2 going in the same direction (see Figure 16). Linearity of the map means that if a point C goes into the point D, then

$$\frac{OC}{OA} = \frac{OD}{OB}.$$

If, under the same transformation, a point E goes into the point F, the similarity of the triangles OCE and ODF (see Figure 16) implies that the

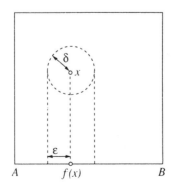

FIGURE 15

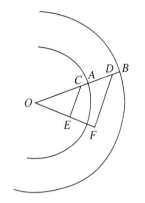

FIGURE 16

length of the interval CE is increased by the factor of R/r. Therefore, a circle goes into a circle, and a disc of radius δ goes into the disc of radius $\varepsilon = (R/r)\delta$. Thus, for any $\varepsilon > 0$ there is the corresponding $\delta = (r/R)\varepsilon$, which means that the mapping is continuous.

Example 3. Let K be a disc of radius R with the center at a point O. Consider the central projection of the disc K onto its boundary circle C. It is defined in the following way. Let x be a point of the disc other than its center. Then $f(x)$ is defined as the point at which the ray issuing from the center O and passing through the point x intersects the circle C (see Figure 17). No mapping is defined for the point O in that way, but we shall assume that the center is projected, for example, into the topmost point A of the circle C. We now show that this projection is discontinuous at the center of the disc and continuous at any other point of the disc. Take an ε-neighborhood of the point $A = f(O)$ on the circle C, sufficiently small so that it does not contain the entire circle. One can, for example, take any

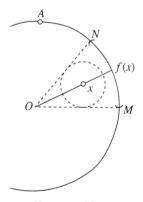

FIGURE 17

$\varepsilon < 2R$. There is no corresponding δ for this value of ε. However small a δ-neighborhood of the point O is, the points of this neighborhood are mapped into points scattered all over the circle, and hence cannot possibly fit into the chosen ε-neighborhood of the point O. Thus, the mapping is discontinuous at the center of the disc.

Let x be a point on the disc other than its center (see Figure 17), $f(x)$ its image, and the arc MN (without the endpoints) the ε-neighborhood of the point $f(x)$. Then the corresponding δ-neighborhood of the point x can be easily found in the following way: take the largest disc with its center at x lying in the circular sector subtended by the arc MN. The radius of this circle is δ. One can derive a formula expressing the dependence of δ on ε (see Problem 14).

We say $f : X \to Y$ is a *one-to-one* mapping if it takes different points of set X into different points of set Y. A mapping $f : X \to Y$ is said to be a *surjection* if each point of set Y is an image of some point (or points) of set X. Thus, under a surjection, set X is mapped on the entire set Y, not just part of it. For that reason a surjection is sometimes referred to as an *onto* mapping.

For a one-to-one surjection $f : X \to Y$ we define the *inverse* mapping $f^{-1} : Y \to X$ associating to each point y in Y the point x in X that is mapped by f into y.

A one-to-one correspondence $f : X \to Y$ is said to be a *homeomorphism* if it is continuous and the inverse mapping f^{-1} is also continuous. Sets X and Y are then said to be *homeomorphic*. A property of a set preserved under any homeomorphism is a *topological* property. Properties of this kind are studied by the science of *topology*.

It follows from Example 2 that any two discs are homeomorphic (try to carry out the proof yourself!). One can verify that any disc is homeomorphic to a square, a triangle, or a polygon. The boundaries of these figures are also homeomorphic (i.e., a circle is homeomorphic to the boundary of a square or a triangle).

A set X is said to have a *fixed point property* if each continuous mapping $f : X \to X$ of this set into itself has a fixed point. We can easily see that this property is a topological one. That is, if a set X has it, then so does any set Y homeomorphic to X. To check this, let $g : X \to Y$ be any homeomorphism, and let $f : Y \to Y$ be any continuous mapping. Consider the mapping $\varphi = g^{-1}fg$ of the set X into itself. This mapping (which is called the *composition* of the mappings g, f, and g^{-1}) is defined in the following way. First, the map g is applied to a point x in X. The result is the point $y = g(x)$ in Y. Then f is applied to y, yielding the point $f(y) = fg(x)$ in Y. Finally, after the application of g^{-1} one gets the point $g^{-1}fg(x)$ in X. One can verify that the mapping φ of the set X into itself is continuous. Therefore, it has a fixed point x_0 in X. Denote the image of this point under the homeomorphism g by y_0, i.e., $y_0 = g(x_0)$. We have

$$\varphi(x_0) = g^{-1}fg(x_0) = x_0.$$

Applying g to both sides of this equality, we get

$$gg^{-1}fg(x_0) = g(x_0),$$

or

(5.1) $$fg(x_0) = g(x_0).$$

Because $g(x_0) = y_0$, formula (5.1) implies that $f(y_0) = y_0$, so that y_0 is a fixed point.

The importance of the statement we have just proved cannot be overestimated. Now we can always verify the fixed point property for just one representative of an entire class of homeomorphic sets. For example, in Chapter 8 this property will be proved for a square, and thereby it will be automatically verified for a disc, triangle, or any convex plane polygon whatsoever.

Problems. 13. Let X and Y be two sets (on a plane), and let f be a mapping from X to Y. What can be said about this mapping if

(a) for any $\delta > 0$ and any $\varepsilon > 0$ the image of each point from the δ-neighborhood of some point x_0 lies in the ε-neighborhood of the point $y_0 = f(x_0)$;

(b) for any $\delta > 0$ there exists $\varepsilon > 0$ such that the image of each point in the δ-neighborhood of the point x_0 lies in the ε-neighborhood of the point $y_0 = f(x_0)$?

14. Prove that the radial projection of a disc on its boundary circle is continuous at any point other than its center. Find the formula expressing the dependence of δ on ε.

15. A *closed interval* $[a, b]$ is said to be *nondegenerate* if $a \neq b$. Prove that all nondegenerate closed intervals are homeomorphic.

16. Prove that the composition of two (or three) continuous mappings is a continuous mapping.

6

Compactness

Let $\{x_1, x_2, \ldots, x_n, \ldots\}$ be a sequence of points on a line, on a plane, or in a three-dimensional space. We shall denote such a sequence by $\{x_n\}$. The individual element of the sequence numbered with n will be denoted x_n. A sequence $\{x_n\}$ is said to *converge to a point* x_0 as n tends to infinity if for any $\varepsilon > 0$ there exists an integer m such that for all $n > m$ the distance between the points x_0 and x_n is less than ε. It is denoted $x_n \to x_0$. In other words, the convergence $x_n \to x_0$ means that for any ε-neighborhood of the point x_0 all the points of the sequence after a certain one lie in this neighborhood. For example, the sequence of points $x_n = (1 + (-1)^n)/n$ on a line converges to zero since the odd members of the sequence are equal to zero, while the member numbered by any even number $n = 2k$ is of the form $x_n = 2/2k = 1/k$. For the inequality $1/k < \varepsilon$ to be satisfied, we need $k > 1/\varepsilon$ whence $n/2 > 1/\varepsilon$, or $n > 2/\varepsilon$. Thus, for any integer n greater than $m = [2/\varepsilon] + 1$ (where the square brackets denote the greatest integer in $2/\varepsilon$) the point x_n numbered by n belongs to the ε-neighborhood of zero. On the other hand, the sequence

$$\left\{ 1, 1 - \frac{1}{2}, \frac{1}{3}, 1 - \frac{1}{4}, \ldots, \frac{1}{2k-1}, 1 - \frac{1}{2k}, \ldots \right\}$$

does not converge, since it includes two subsequences converging to different points. Indeed, its subsequence

$$\left\{ 1, \frac{1}{3}, \frac{1}{5}, \ldots, \frac{1}{2k-1}, \ldots \right\}$$

converges to zero, while the subsequence

$$\left\{ 1 - 1/2, 1 - 1/4, 1 - 1/6, \ldots, 1 - 1/2k, \ldots \right\}$$

converges to 1.

With the help of sequences the notion of compactness, one of the most important in topology, is defined.

A set X is said to be *compact* if any sequence $\{x_1, x_2, \ldots, x_n, \ldots\}$ of points of X contains a subsequence $\{x_{n_1}, x_{n_2}, \ldots, x_{n_k}, \ldots\}$ converging to some point x of the same set X.

For example, any finite set is compact because any sequence of its points contains a stationary subsequence, all the points of which are equal, and such a sequence evidently converges. We shall now prove that both a closed interval and a square are compact. All convex polygons have the same property, as well as a disc, a circle, and a sphere.

Let us start with two examples of noncompact sets. They are the real line $\mathbf{R}$ and its open interval $(0, 1)$. Indeed, the sequence of natural numbers $\{1, 2, 3, \dots, n, \dots\}$ in $\mathbf{R}$ has no convergent subsequence, so that the real line $\mathbf{R}$ is not compact. On the other hand, while the sequence $\{1/2, 1/3, 1/4, \dots, 1/n, \dots\}$ of points of the interval $(0, 1)$ converges to zero, its limit (zero) does not belong to this interval whereby the compactness condition is again violated.

The proof of compactness of a closed interval is based on the Principle of Nested Closed Intervals. This principle (or an equivalent statement) is usually taken as one of the axioms of the real line.

PRINCIPLE OF NESTED CLOSED INTERVALS. *If a sequence of closed intervals $\{I_0, I_1, \dots, I_n, \dots\}$ is such that each interval lies inside the preceding one, and their lengths tend to zero as n tends to infinity, then all the closed intervals have a unique common point.*

Let us prove that the closed interval $I_0 = [a, b]$ is compact.

Let $\{x_1, x_2, \dots, x_n, \dots\}$ be any sequence of points lying in $[a, b]$. Divide the interval $[a, b]$ in half. At least one of the halves contains infinitely many (either distinct or coinciding) points of the sequence $\{x_n\}$. Denote this half by I_1, and choose a point x_{n_1} of the sequence $\{x_n\}$ in it. Divide the closed interval I_1 also in half, and denote by I_2 the half that contains infinitely many points of the sequence $\{x_n\}$. Proceeding in the same way, we get the sequence of the "nested" closed intervals $\{I_0 = [a, b], I_1, \dots, I_k, \dots\}$. The length of the closed interval I_k equals $(b - a)/2^k$, and the closed interval I_k contains a point x_{n_k} of the sequence $\{x_n\}$. The lengths of the closed intervals I_k evidently tend to zero. According to the Principle of Nested Closed Intervals there is a unique point x_0 belonging to all intervals I_k $(k = 0, 1, 2, \dots)$, and to the interval $[a, b]$ in particular. Let us prove that the subsequence of points $\{x_{n_1}, x_{n_2}, \dots, x_{n_k}, \dots\}$ selected from the sequence $\{x_n\}$ tends to x_0. Take an arbitrary $\varepsilon > 0$. Then there exists an integer m such that the length of the closed interval I_m (equal to $(b - a)/2^m$) is less than ε. The closed interval I_m contains the point x_0, and therefore lies inside the ε-neighborhood of this point. The same is true for all closed intervals I_k with indices $k \geq m$. Thus, the ε-neighborhood of the point x_0 contains all points of the subsequence $\{x_{n_k}\}$ starting with the index n_m. Therefore, the sequence $\{x_{n_k}\}$ converges to the point x_0, which proves that the closed interval I_0 is compact.

Now we prove the compactness of a square. Let Q_0 be a square whose sides are parallel to the coordinate axes Ox and Oy and equal to 1. Let $\{p_n\}$ be an arbitrary sequence of points in Q_0. Cut Q_0 into four equal squares

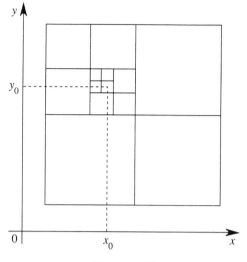

FIGURE 18

with straight lines parallel to its sides. Denote by Q_1 a smaller square (one of these four) that contains infinitely many (either distinct or coinciding) points of the sequence $\{p_n\}$. Choose one of these points and denote it by p_{n_1}. Cut the square Q_1 again into four equal squares, and denote by Q_2 the square that contains infinitely many points of the sequence $\{p_n\}$. Choose one of them and denote it by p_{n_2}. Proceeding in the same way, we get the sequence of nested squares $\{Q_0, Q_1, \ldots, Q_k, \ldots\}$ such that the side length of the square Q_k equals $1/2^k$, and Q_k contains the point p_{n_k} of the sequence $\{p_n\}$ (see Figure 18). The projections of these squares on the Ox axis forms a nested system of closed intervals $\{I_0^x, I_1^x, \ldots I_k^x, \ldots\}$. Let x_0 be the unique common point of these intervals. Now let $\{I_0^y, I_1^y, \ldots I_k^y, \ldots\}$ be the sequence of nested closed intervals on the Oy axis that are the projections of the squares $\{Q_0, Q_1, \ldots, Q_k, \ldots\}$ on it. Denote by y_0 the unique common point of these intervals. We leave it to the reader to prove that the subsequence $\{p_{n_k}\}$ converges to a point of the square Q_0 with coordinates (x_0, y_0). This concludes the proof that the square Q_0 is compact.

Problem. 17. Prove that if a function f is continuous on the closed interval $[a, b]$, then it is bounded from both above and below. That is, there are numbers m and M such that $m \leq f(x) \leq M$ for all points x of the closed interval.

7

Proof of Brouwer's Theorem for a Closed Interval, the Intermediate Value Theorem, and Applications

Brouwer's theorem for a closed interval has been formulated in Chapter 1. We shall now give its proof using the rigorous notion of continuity in the "$\varepsilon - \delta$ language."

The basic idea of the proof is to divide the closed interval into subintervals and look for a subinterval whose endpoints are moved by our mapping in opposite directions. These moves must necessarily be small, otherwise the mapping would be discontinuous. In other words, the subinterval can be considered as an approximate fixed "point." The real fixed point is then "found" explicitly using the Principle of Nested Closed Intervals.

Let f be a continuous mapping of a closed interval $[a, b]$ into itself. If at least one of the endpoints of $[a, b]$ stays where it was, the theorem is proved. So let us assume that the mapping shifts both endpoints, point a to the right, and point b to the left. Divide the interval $I_0 = [a, b]$ in half by the midpoint $c = (a + b)/2$. If point c stays where it was, the theorem is proved, otherwise it moves either right or left. In both cases one of the halves of the closed interval I_0 has the endpoints that are shifted by the mapping f inside the corresponding interval. Denote this half by I_1. Repeat the process by dividing the interval I_1 in half. Then either its midpoint is a fixed point, or there is a half I_2 of the interval I_1 with endpoints that are shifted inside it. Proceeding in the same way, either we find a fixed point in a finite number of steps or obtain an infinite sequence $\{I_0, I_1, I_2, \ldots, I_n, \ldots\}$ of nested closed intervals such that the function f shifts their endpoints inside the respective intervals.

Let x_0 be the unique common point of these intervals. Let us prove that $f(x_0) = x_0$. This is a direct consequence of the continuity condition. To be more precise, suppose that the points x_0 and $y_0 = f(x_0)$ do not coincide, and that y_0 lies to the right of x_0. The mapping f is continuous at the point x_0. In the $\varepsilon - \delta$ language this means that for any ε-neighborhood of the point y_0 there is a δ-neighborhood of the point x_0 such that images of all points of it belong to the ε-neighborhood of the point y_0. The choice

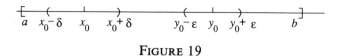

FIGURE 19

of $\varepsilon > 0$ (the radius of the neighborhood) is up to us. Choose it sufficiently small for the ε- and δ-neighborhoods to be disjoint (see Figure 19).

Because all the points of the chosen δ-neighborhood are mapped into the ε-neighborhood, the function shifts all of them to the right. For a sufficiently large n the entire closed interval I_n lies inside the δ-neighborhood of the point x_0, and therefore both endpoints move to the right, contradicting the fact that the left endpoint of I_n moves to the right while its right end moves to the left. The proof is completed.

As we explained in the Preface, the problem of finding solutions for the equation $f(x) = 0$ in the closed interval $[a, b]$ can be reduced to the problem of finding fixed points of the mapping F where $F(x) = \lambda f(x) + x$. Consider this situation in more detail, and let us first prove the following theorem.

INTERMEDIATE VALUE THEOREM. *Let f be a continuous function defined in the closed interval $[a, b]$ such that $f(a) < f(b)$. If a number c satisfies the condition $f(a) < c < f(b)$, then there is a point x_0 in the interval for which $f(x_0) = c$.*

In other words, a continuous function in a closed interval takes all intermediate values.

First, we shall prove the theorem in a special case when $c = 0$, and, consequently, $f(a) < 0$ and $f(b) > 0$. The idea of the proof is, starting from the function $f(x)$ on $[a, b]$, to construct a new continuous function $F(x) = \lambda f(x) + x$ defined in the same closed interval where the parameter $\lambda \neq 0$ is chosen in such a way that the function $F(x)$ maps the closed interval $[a, b]$ into itself. This choice ensures that the function $F(x)$ has a fixed point x_0 in the closed interval $[a, b]$. Then $\lambda f(x_0) + x_0 = x_0$, and, therefore, $f(x_0) = 0$. The process of choosing λ is shown in Figure 20. The function $F(x)$ maps the closed interval $[a, b]$ into itself if its graph lies inside the square $ABCD$. Therefore, the coefficient λ must be sufficiently small in absolute value. In other words, the graph of $f(x)$ must be contracted. Moreover, λ must be negative because otherwise any contraction takes the graph of $F(x)$ out of the square.

Therefore, the problem is to determine the precise value of λ. We will find it in the following way. The continuous function $f(x)$ is bounded on $[a, b]$. That is, there exist m and M such that $m \leq f(x) \leq M$ for any point x of the interval (see Problem 17). In our case one evidently has $m < 0$ and $M > 0$. Since $f(a) < 0$, the continuity of the function $f(x)$ implies that $f(x) < 0$ for all x sufficiently close to a. Choose a point x_1 such that $f(x) < 0$ for all $a \leq x \leq x_1$. Next, choose a point x_2 such that

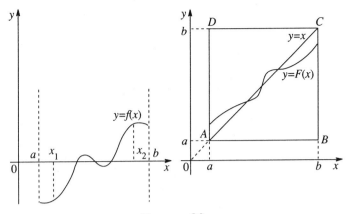

FIGURE 20

$f(x) > 0$ for all $x_2 \leq x \leq b$ (see Figure 20). Now let

$$\lambda = \max \left\{ \frac{a - x_1}{M}, \frac{b - x_2}{m} \right\},$$

that is, λ is equal to one of the two negative numbers $(a - x_1)/M$ or $(a - x_2)/m$, namely, the one that is not the least (not the greatest in absolute value).

We now show that for such λ one has $F(x) \geq a$ if $a \leq x \leq b$. Let us first consider points x for which $f(x) \geq 0$. Take the inequality $\lambda \geq (a - x_1)/Mf(x)$ and multiply both its sides by $f(x)$. Then $\lambda f(x) \geq (a - x_1)/Mf(x)$, and, using the inequality $-f(x) \geq -M$, we have

$$F(x) = \lambda f(x) + x \geq \frac{a - x_1}{M} f(x) + x$$

$$= \frac{a - x_1}{(-M)}(-f(x)) + x \geq \frac{a - x_1}{(-M)}(-M) + x = a - x_1 + x.$$

We have $f(x) \geq 0$, so that the choice of the point x_1 implies that $x > x_1$, and hence $F(x) \geq a - x_1 + x > a$. Consider now the point x for which $f(x) < 0$. Then $\lambda f(x) > 0$ and hence

$$F(x) = \lambda f(x) + x > x \geq a.$$

Let us show now that if $a \geq x \geq b$ then $F(x) \leq b$. Consider a point x for which $f(x) \geq 0$. Then $\lambda f(x) \leq 0$, and therefore

$$F(x) = \lambda f(x) + x \leq x \leq b.$$

For a point x such that $f(x) < 0$, take the inequality $\lambda \geq (b - x_2)/mf(x)$ and multiply both sides by $f(x)$. Then $\lambda f(x) \leq ((b - x_2)/m)f(x)$, and using the inequality $-f(x) \leq -m$, we have

$$F(x) = \lambda f(x) + x \leq \frac{b - x_2}{m} f(x) + x = \frac{b - x_2}{(-m)}(-f(x)) + x$$

$$\leq \frac{b - x_2}{(-m)}(-m) + x = b - x_2 + x.$$

Because $f(x) < 0$, the choice of the point x_2 implies that $x < x_2$, hence $F(x) \leq b - x_2 + x < b$.

Thus, the continuous function $F(x)$ maps the closed interval $[a, b]$ into itself and therefore has a fixed point x_0 for which $f(x_0) = 0$.

In the case when $c \neq 0$ the argument is the same, except that the function $f(x)$ is replaced with $f(x) - c$. In particular, we now have $F(x) = \lambda[f(x) - c] + x$. A detailed proof is left to the reader.

The intermediate value theorem is evidently true also in the case where $f(a) > f(b)$ and the number c satisfies the inequalities $f(a) > c > f(b)$. We suggest the reader to try to carry out the proof in this case as well.

Finally, note that the fixed point theorem is a consequence of the intermediate value theorem (Problem 18).

Problems. 18. Derive the fixed point theorem for a closed interval from the intermediate value theorem.

19. Give a proof of the intermediate value theorem independent of the fixed point theorem.

20. Let f be a continuous function on the closed interval $[a, b]$ that takes each endpoint of the interval into itself, that is, $f(a) = a$ and $f(b) = b$. Let g be any continuous function taking the closed interval $[a, b]$ into itself. Prove there is a point x_0 in this closed interval such that $f(x_0) = g(x_0)$. Is the same statement valid for an arbitrary continuous function g on $[a, b]$?

21. Prove that a continuous mapping f of a closed interval $A = [a_1, a_2]$ onto the (entire) closed interval $B = [b_1, b_2]$ containing A has a fixed point. Is the statement true if the image $f(A)$ of the closed interval A does not coincide with B, but only lies inside it?

22. Let f be a continuous mapping of the closed interval $[a, b]$ into itself. Prove that the mapping $g = f^2$ (the composition of the mapping f with itself) has at least two fixed points. Is the statement true if f maps the closed interval $[a, b]$ into itself, but not onto itself?

23. The function f defined on the closed interval $[a, b]$ is said to be *nondecreasing* if $x_1 < x_2$ implies $f(x_1) \leq f(x_2)$, and *nonincreasing* if $x_1 < x_2$ implies $f(x_1) \geq f(x_2)$. Prove that any nondecreasing (or nonincreasing) function that maps a closed interval into itself has a fixed point in the closed interval even if it is not necessarily continuous.

24. Prove that the cubic equation

$$a_0 x^3 + a_1 x^2 + a_2 x + a_3 = 0$$

where $a_0 \neq 0$, and the other coefficients are arbitrary real numbers, has at least one real root. For $a_3 = 0$ there is obviously the root $x = 0$. Can we tell if there is a positive or negative root, if $a_3 \neq 0$?

Let us now present some corollaries to the intermediate value theorem. Consider continuous mappings of a circle C into the real line **R**. The position of a point x on the circle is usually described by the angle α formed by the radius drawn from the center of C to x with the radius directed horizontally

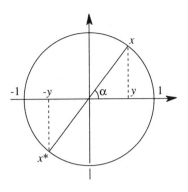

FIGURE 21

to the right of the center (see Figure 21). The angle α is said to be the *angular coordinate* of the point x.

The counterclockwise direction is usually taken for the positive one. We shall assume for the time being that α varies between 0 and 2π (measured in radians). Two points x and x^* lying on the circle are said to be *antipodal* if their angular coordinates differ by π (or if the chord connecting them is a diameter, which is the same).

THE BORSUK-ULAM THEOREM FOR A CIRCLE. *Let* $f : C \to \mathbf{R}$ *be a continuous function defined on a circle. Then there is a pair of antipodal points* x *and* x^* *such that* $f(x) = f(x^*)$.[1]

PROOF. Let C be the unit circle with its center at the origin, and α the angular coordinate of its point x (see Figure 21). Define a continuous function $g(y)$ on the closed interval $I = [-1, 1]$ as follows. For a point y on I, let x be the point of intersection of the perpendicular raised from the point y on the horizontal axis with the upper semicircle, and let x^* be the point antipodal to x. Let

$$g(y) = f(x) - f(x^*) = f(\alpha) - f(\alpha \pm \pi).$$

The values of the function $g(y)$ at the endpoints of the closed interval I are $g(1) = f(0) - f(\pi)$ and $g(-1) = f(\pi) - f(0)$. That is, $g(1) = -g(-1)$. If $g(1) = 0$, the theorem is proved. Otherwise the intermediate value theorem implies that there is a point y in the closed interval I such that $g(y) = 0$. Now if x is the point of the circle corresponding to y then $f(x) = f(x^*)$.

COROLLARY. *At any given moment of time and on any great circle of the Earth (the equator, for example) there is a pair of antipodal points where the air temperature is the same.*

FIRST PANCAKE THEOREM. *If* A *and* B *are two bounded plane figures, then there is a line dividing each into two parts of equal area.*

---

[1]K. Borsuk (1905–1982) was a Polish mathematician, and S. Ulam (1909–1984) was an American mathematician of Polish origin.

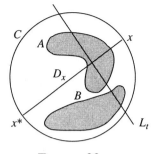

FIGURE 22

This theorem assumes each of the two figures have an area (which is not necessarily the case, e.g. if a figure has a very complicated and intricate form). The significance of the theorem's name is now clear. Two pancakes served on the same plate can be cut exactly in half with a single slash of a knife.

For the proof let us put both figures inside a circle C (see Figure 22). (We can do this because both figures are bounded.).

For any point x on the circle C denote by D_x the diameter of C passing through x. Let L_t be perpendicular to D_x, passing through the point on D_x located at the distance t from x ($0 \le t \le d$, where d is the diameter of the circle). Let $f_1(t)$ be the area of figure A that lies on the same side of L_t as x, and let $f_2(t)$ be the area of the other part. The functions $f_1(t)$ and $f_2(t)$ are defined in the closed interval $[0, d]$, and both are continuous on it, together with their difference $f(t) = f_1(t) - f_2(t)$. Evidently, one has $f(0) = -f(d)$, and therefore there is a point t in the closed interval $[0, d]$ such that $f(t) = 0$ or $f_1(t) = f_2(t)$. Thus, the line L_t passing through this point cuts A in half. This line also cuts B into two parts. Denote by $g_1(x)$ the area of that part that lies closer to the point x, and by $g_2(x)$ the area of the other part. Consider the function $g(x) = g_1(x) - g_2(x)$. It is defined on the circle C and is continuous on it. When the point x, moving continuously along C, comes into the antipodal point x^*, the two parts of the figure B change places. Hence $g(x) = -g(x^*)$ for all points x of the circle C. By the Borsuk-Ulam theorem there is a point x at which $g(x) = g(x^*)$. The equalities $g(x) = -g(x^*)$ and $g(x) = g(x^*)$ imply that $g(x) = 0$. That is, there is a point x on the circle for which line L_t cuts both figures in half.

SECOND PANCAKE THEOREM. *If A is a bounded plane figure then there exist two perpendicular lines cutting A into four parts of equal area.*

As in the previous case, place the figure A inside a circle C. For any point x on C denote by D_x the diameter passing through the antipodal points x and x^*. Now let L_x be the straight line perpendicular to D_x that also cuts A in half and let M_x be the line also cutting A in half but parallel to D_x (the existence of both lines has already been proved). These two lines cut A into four parts, which we shall denote counterclockwise by $A_1(x)$,

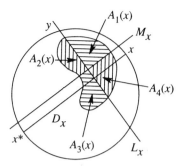

FIGURE 23

$A_2(x)$, $A_3(x)$, and $A_4(x)$ (see Figure 23). Let $S_1(x)$, $S_2(x)$, $S_3(x)$, and $S_4(x)$ be their respective areas. Evidently, one has

$$S_1(x) + S_2(x) = S_3(x) + S_4(x)$$

and

$$S_4(x) + S_1(x) = S_2(x) + S_3(x),$$

whence $S_1(x) = S_3(x)$ and $S_2(x) = S_4(x)$.

These equalities have been obtained for an arbitrary but fixed position of the point x. Now suppose this point moves along the circle C counterclockwise continuously and covers the circular arc equal to $\pi/2$ radian. Denote its new position by y. Then the part $A_1(x)$ of the figure A continuously transforms into $A_2(x)$, ... , and the part $A_4(x)$ transforms into $A_1(x)$. Hence $S_1(y) = S_2(x)$ and $S_2(y) = S_3(x)$.

Let $f(x) = S_1(x) - S_2(x)$. Then

$$f(y) = S_1(y) - S_2(y) = S_2(x) - S_3(x) = S_2(x) - S_1(x) = -f(x).$$

Therefore, the continuous function f changes its sign as x covers the circular arc of $\pi/2$ radian. Now a usual argument shows that $f(x) = 0$ at some point x of the arc. This point gives the desired decomposition of the figure A, and the proof is completed.

Note that in both pancake theorems we did not check the continuity of the functions arising there. For the proof that they are indeed continuous see [5, §11].

We shall conclude this chapter with a problem from mechanics. A train moves along a straight section of a railroad track between two stations. In one of the cars a rod is attached to the floor by a hinge. Suppose that the axis of the hinge is parallel to the axes of the car wheels. The rod is left to its fate and can take any position between the two extreme ones (lying on the floor facing forward or backward). Let us ask the question: is there an initial position of the rod ensuring that it never touches the floor of the car during the whole journey? We will find that the answer is yes. To get this answer neither precise knowledge of the laws of dynamics nor the analysis of forces acting on the rod is necessary. It is sufficient to make only one assumption related to physics, namely that the final position of the rod depends on its

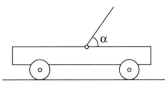

FIGURE 24

initial position continuously. In particular, if the rod begins the journey lying on the floor, it will stay in this position throughout the journey.

The position of the rod at each moment of time is determined by a single coordinate, the angle α it forms with the floor (see Figure 24).

The angles $\alpha = 0$ and $\alpha = \pi$ correspond to the two opposite extreme positions of the rod when it lies on the floor. Let us assume that the statement is not true, i.e., that under any initial conditions the rod falls either forward or backward so that α eventually assumes the value of either 0 or π and then remains constant. Define a function $f(\alpha)$ on the closed interval $[0, \pi]$ that associates to each initial position of the rod its final position. In other words, if, for example, the initial value of the angle α was $\pi/6$, and the rod eventually fell to the right, one sets $f(\pi/6) = 0$.

According to our assumption, the function $f(\alpha)$ is continuous on the closed interval $[0, \pi]$ and assumes the following values at its endpoints: $f(0) = 0$, $f(\pi) = \pi$. Moreover, *each* value the function assumes is either 0 or π. Because this clearly contradicts the intermediate value theorem, the statement is proved.

The above proof is a "pure existence proof" providing no indication of how the desired initial position can actually be found. The same applies to the pancake cutting problems as well.

Problems. 25. Prove that a circle is not homeomorphic to the straight line or to any subset of the straight line.

26. Find a mistake in the following argument. The air temperature at any particular place, say the city of Sverdlovsk, changes continuously with time. Let us follow its change more closely during an hour, while the long hand of a clock makes a round trip. According to the Borsuk-Ulam theorem there are two moments of time differing by half an hour at which the air temperature is the same. The clock's hand, however, can be assumed to move at any speed, while the beginning and end of its motion can be chosen arbitrarily. Hence the air temperature is the same at any two moments of time, and hereby constant.

27. It is evident that if one pancake has the form of a disc, and the other that of a square, then the line cutting both pancakes in half goes through their centers. Is this statement true for any two pancakes in the shape of regular polygons?

28. Prove that a square can be circumscribed around any closed plane curve.

8

Proof of Brouwer's
Theorem for a Square

We shall give two proofs of this theorem. Each of them is directed to its "own" reader. For those who want to understand the basic ideas without too much detail, we provide a clear-cut but "not entirely rigorous" proof on the chessboard, taken from the book [6, Russian p. 141]. For those who are not satisfied with "approximate" arguments we will give the rigorous proof using the $\varepsilon-\delta$-language. Incidentally, the basic idea in both cases is the same. As in the one-dimensional case, one looks for a subsquare (instead of a subinterval) whose points move in different directions. Then in order to satisfy the continuity condition, these displacements must be considered small or vanishing, thereby indicating the presence of a fixed point.

Consider a continuous mapping of a square (on a chessboard) into itself. For each point p of the square denote by q its image under the mapping. Let us call a square on the chessboard *red* if for any point p the corresponding point q lies closer to the right edge of the board than the point p itself. If for each point p of the square its image q lies closer to the left edge of the board than the point p, the square will be called *blue*. Squares that are neither red nor blue will be called *yellow*. The image of a point lying on the right edge of the chessboard can not be closer to the edge than the point itself. Therefore, no square in the extreme right column is red. Similarly, no square in the extreme left column is blue. No point can move to the right and to the left at the same time, and therefore no red square will border a blue one (since neighboring squares have at least one point in common). Thus, if a king is not allowed to enter a yellow square, it will never be able to cross the chessboard from its extreme left to its extreme right column, which implies that a rook can cross the board from bottom to top moving over yellow squares only (see Problem 7). Therefore, one can draw an uninterrupted polygonal line from the bottom to the top going through the yellow squares only. For example, such is the line connecting centers of successive squares passed by the rook.

Take such a line and draw an arrow from each point p on it to the corresponding point q (the displacement vector). At the initial point of the line lying on the bottom edge of the chessboard the (p, q) arrow cannot be directed downward (even if it is also inclined to the right or to the left). It has to be directed upwards. In the last point of the line lying on the top edge

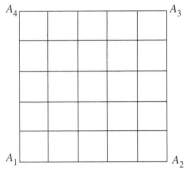

FIGURE 25

of the board, the (p, q) arrow looks, generally speaking, downward. When one moves along the line, the direction of the arrow varies continuously, so there is a point p_1 on the line where the arrow (p_2, q_2) is horizontal. The definition of a yellow square implies that in the square containing the point p_1 there is a point p_2 where the arrow (p_2, q_2) is vertical. If the square is small, such a jump in the direction of the arrow is possible only if arrows themselves are small for all points p of this square.

Dividing the board into n^2 squares and letting n tend to infinity, we obtain the point p_0 at which the arrow (p_0, q_0) vanishes, i.e., a fixed point.

Now we present a rigorous proof. Consider a square $Q = A_1 A_2 A_3 A_4$ and let us prove that any continuous mapping f of this square into itself has a fixed point.

Assume a square Q is subdivided into smaller squares ("faces") by straight lines parallel to its sides (see Figure 25).

If some vertex of the subdivision remains unmoved by the mapping f, the theorem is proved. Assume, therefore, that all vertices are displaced. We label each vertex of the subdivision by 1, 2, 3, or 4. The labels are assigned according to the direction of the displacement in such a way that conditions of Lemma 4 (see p. 11) are satisfied. More precisely, let p be a vertex of the subdivision, let $q = f(p)$ be its image under the mapping f, let (p, q) be the displacement vector, and let φ be the angle it forms with the positive direction of the horizontal axis. The vertex p of the subdivision is then labeled according to Table 1. (See also Figure 26.)

In particular, if the point p coincides with the vertex A_1 of the square Q, the angle φ satisfies the inequalities $0 \leq \varphi \leq \pi/2$. When the strict inequalities $0 < \varphi < \pi/2$ hold, the vertex A_1 is, according to Table 1, labeled by 1. However, if $\varphi = 0$ or $\varphi = \pi/2$ then, as the first and the third rows of the table show, this vertex can be assigned label 1, 2, or 4. In order to satisfy the conditions of Lemma 4, we choose one label, namely 1. Similarly, vertex A_2 is labeled by 2, vertex A_3 is labeled by 3, and vertex A_4 is labeled by 4.

If the point p belongs to the side $A_1 A_2$ of the square Q and does not coincide with its endpoints, then $0 \leq \varphi \leq \pi$. If the strict inequalities $0 <$

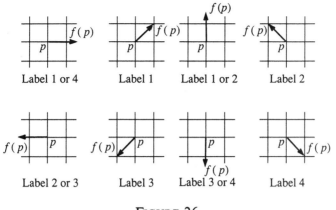

FIGURE 26

$\varphi < \pi/2$ or $\pi/2 < \varphi < \pi$ are satisfied, then according to Table 1 the point $\varphi = 0$ is labeled by 1 or 2, respectively. If $\varphi = 0$ at this point, we can choose between labels 1 or 4. We choose 1 in order to satisfy the conditions of Lemma 4. The same procedure is applied if $\varphi = \pi/2$ or $\varphi = \pi$.

Similar arguments apply to points of the three remaining sides of the square. Finally, the points inside the square are labeled strictly according to Table 1, without regard to Lemma 4.

The system of labels obtained in this way satisfies conditions of Lemma 4. Using this lemma we conclude that our subdivision has a face that carries at least three different labels.

Now consider a sequence of decompositions $\{\tau_1, \tau_2, \ldots, \tau_n, \ldots\}$ of the square Q. The decomposition τ_n with the index n is constructed as follows. Each side of the square is divided into 2^n equal parts, and the straight lines are drawn through these points parallel to its sides. Clearly, the length of all edges of the subdivision τ_n tends to zero as n tends to infinity.

Table 1

Angle	Mark
$\varphi = 0$	1 or 4
$0 < \varphi < \frac{\pi}{2}$	1
$\varphi = \frac{\pi}{2}$	1 or 2
$\frac{\pi}{2} < \varphi < \pi$	2
$\varphi = \pi$	2 or 3
$\pi < \varphi < \frac{3\pi}{2}$	3
$\varphi = \frac{3\pi}{2}$	3 or 4
$\frac{3\pi}{2} < \varphi < 2\pi$	4

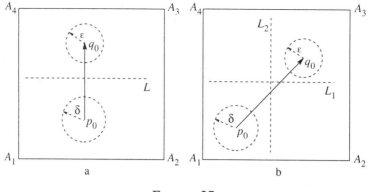

FIGURE 27

Suppose all vertices of each decomposition τ_n are displaced under the mapping f. Then there is a face (subsquare) in each decomposition labeled by at least three different numbers. Choose one such square in each subdivision τ_n ($n = 1, 2, \ldots$) and denote it by Q_n. Let x_n, y_n, z_n, u_n be its vertices.

Because the square Q is compact, one can select a subsequence of the sequence $\{x_1, x_2, \ldots, x_n, \ldots\}$ converging to a point p_0 in Q. We can assume as before that it is the sequence $\{x_n\}$ itself. Then there are three more sequences that converge to the same point, namely $\{y_n\}, \{z_n\}, \{u_n\}$, because the length of the side of the square Q_n tends to zero as n tends to infinity.

Let us prove that p_0 is a fixed point. Suppose points p_0 and $q_0 = f(p_0)$ are distinct, and consider two cases. First, let point q_0 lie strictly above point p_0. In other words, $\varphi = \pi/2$ for p_0 (see Figure 27a). (The angle φ can be defined even for points p_0 that are not vertices of the subdivision). Draw a horizontal straight line L separating points p_0 and q_0. Take an ε-neighborhood of point q_0 satisfying only the condition that it has no common points with L. Since the mapping f is continuous at point p_0, there is a δ-neighborhood of p_0 such that the images of all points from this neighborhood lie in our ε-neighborhood of q_0. We will choose this δ-neighborhood to be small enough to have no common points with line L.

Then the δ-neighborhoods and the ε-neighborhood not only have no points in common, but any point of the ε-neighborhood of point q_0 lies above any point of the δ-neighborhood of point p_0. Hence, if p is any point (such as a vertex of the subdivision) in the δ-neighborhood of point p_0 and $q = f(p)$ is its image, then the displacement vector (p, q) is directed upwards with a possible tilt to the left or right. In other words, the angle φ corresponding to point p satisfies the inequalities $0 < \varphi < \pi$. Therefore, such a point must be labeled by 1 or 2.

On the other hand, for n sufficiently large, the vertices x_n, y_n, z_n, u_n of the square Q_n lie inside the chosen δ-neighborhood of point p_0, so that none of them can be labeled by any number other than 1 or 2. This contradicts the fact that the vertices of Q_n are labeled by at least three different numbers.

We now consider the second case, where the displacement vector (p_0, q_0), looking upwards, tilts right so that one has $0 < \varphi < \pi/2$ for point p_0. (See Figure 27b.) Draw two lines: L_1 horizontal and L_2 vertical in such a way that each of them separates the points p_0 and q_0. Choose an ε-neighborhood of point q_0 and a δ-neighborhood of point p_0 as before, but subject to the condition that neither of them has a common point with L_1 or L_2. If p is any point of the δ-neighborhood of point p_0, and $q = f(p)$, then the angle φ corresponding to the point p satisfies the inequalities $0 < \varphi < \pi/2$. Hence such a point may have only one label, namely 1.

If one now considers, instead of point p, the vertices of the square Q_n for a sufficiently large n, this again results in a contradiction.

We have considered two cases corresponding to the first two rows of the table. The remaining six cases are left to the reader. Brouwer's theorem for a square is proved.

9

The Iteration Method

Let f be a continuous mapping of the closed interval $[a, b]$ into itself. Consider the problem of solving the equation

$$(9.1) \qquad\qquad x = f(x),$$

approximately. In other words, we want to find an approximation to a fixed point of the mapping f. We can try to solve equation (9.1) as follows. Take for the "initial approximation" an arbitrary number x_1 inside the closed interval $[a, b]$ and substitute it into the right hand side of the equation. The value $x_2 = f(x_1)$ is taken for the second approximation. In general, for each successive approximation x_n the next one, x_{n+1}, is found by the formula $x_{n+1} = f(x_n)$.

If a sequence of numbers $\{x_n\}$ (called the *iteration sequence*) has a limit x_0, then x_0 is an (exact) root of equation (9.1). Indeed, if n tends to infinity, the left side of the equality $x_{n+1} = f(x_n)$ tends to x_0, while the right side tends to $f(x_0)$, whence $x_0 = f(x_0)$.

In practice, any process of finding *iterations* x_n is terminated after a finite number of steps, namely, when one has $x_n \approx x_{n+1}$ up to the chosen accuracy. Since $x_{n+1} = f(x_n)$, this means that, up to the chosen accuracy, $x_n \approx f(x_n)$, i.e., x_n is an approximation to the desired root.

Such a scheme of finding an approximate solution of equation (9.1) is called the *iteration method* or the *method of successive approximations*.

The iteration method has a convenient geometric interpretation. As we know (see Chapter 1) finding a root of equation (9.1) is equivalent to finding the abscissa of the common point of the graph of the function $y = f(x)$ and the straight line $y = x$. Figure 28 on p. 40 shows both graphs together with the so-called *iteration path*. This path is constructed as follows. Draw the vertical line through point x_1 on the Ox axis up to point A_1 where it intersects the graph of $y = f(x)$. The coordinates of A_1 are $(x_1, x_2) = (x_1, f(x_1))$. Next draw the horizontal line through A_1 up to its intersection with the straight line $y = x$ at point B_1. The coordinates of this point are $(x_2, x_2) = (f(x_1), f(x_1))$. Again draw the vertical line through B_1 until it meets the graph of $y = f(x)$ at point A_2 with the coordinates $(x_2, x_3) = (x_2, f(x_2))$, etc.

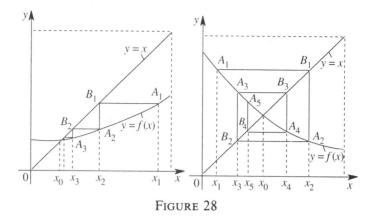

FIGURE 28

Figure 28 shows very clearly how the iteration process converges. The points A_1, A_2, A_3, ... approach the desired intersection point, and their abscissas x_1, x_2, x_3, ... tend to the root of the equation.

We can see easily that an iteration scheme does not necessarily converge in all cases. Consider, for example, the equation $x = f(x)$, where $f(x) = 1-x$. Taking $x_1 = \frac{1}{4}$ for the initial approximation, one has

$$x_2 = x_4 = x_6 = \ldots = \frac{3}{4}, \qquad x_3 = x_5 = x_7 = \ldots = \frac{1}{4},$$

so that the iteration sequence does not converge. The process is said to "go into a loop." The reader will easily verify that in this case the iteration path consists of the four sides of the square and no longer resembles a "staircase," or a "spiral" as in Figure 28.

We now formulate the sufficient conditions for the convergence of an iteration scheme. Let us introduce the following notation. A mapping f of the closed interval $[a, b]$ into itself is said to be a *contraction mapping* (or simply a *contraction*) if it decreases the distance between any two points of this closed interval at least by a factor of M, where $M > 1$. Since the distance between the points x_1 and x_2 equals $|x_1 - x_2|$, the condition can be formulated as follows. There is a number α, such that $0 < \alpha < 1$ and for any two points x_1 and x_2 of the closed interval $[a, b]$ one has

$$(9.2) \qquad\qquad |f(x_1) - f(x_2)| \le \alpha|x_1 - x_2|.$$

The number α is related to the number M by the formula $\alpha = 1/M$. Inequality (9.2) easily implies the continuity of the mapping f.

Let f be a contraction mapping of the closed interval $[a, b]$ into itself. By the intermediate value theorem in its sharpened form (see Problem 31), the image of the closed interval $[a, b]$ is also a closed interval $[a_1, b_1]$ contained in $[a, b]$. The end points of the interval $[a, b]$ *do not necessarily* go into the points a_1 and b_1. Their inverse images are some points x_1 and x_2 in the interval $[a, b]$: $f(x_1) = a_1$, $f(x_2) = b_1$. Since f is a contraction mapping, inequality (9.2) implies that

$$|a_1 - b_1| = |f(x_1) - f(x_2)| \le \alpha|x_1 - x_2| \le \alpha|a - b|.$$

Hence, the mapping sends the closed interval $[a, b]$ into its part $[a_1, b_1]$ of the length not exceeding $\alpha|a - b|$. Then the closed interval $[a_1, b_1]$ is also sent into its part, the interval $[a_2, b_2]$ such that

$$|a_2 - b_2| \leq \alpha|a_1 - b_1|.$$

Let $[a_1, b_1], [a_2, b_2], \ldots, [a_n, b_n], \ldots$ be the system of closed intervals obtained from $[a, b]$ by the successive application of the mapping f. Then

$$|a_n - b_n| \leq \alpha|a_{n-1} - b_{n-1}|,$$

for all n, and so that

(9.3) $$|a_n - b_n| \leq \alpha|a_{n-1} - b_{n-1}| \leq \alpha^2|a_{n-2} - b_{n-2}| \leq \ldots \alpha^n|a - b|.$$

Because $0 < \alpha < 1$, the sequence of numbers $\{\alpha^n\}$ tends to zero. Therefore, the lengths of the closed intervals $[a_n, b_n]$ also tend to zero as n tends to infinity. Thus, there is a unique point x_0 belonging to each of these closed intervals.

Now let x_1 be any other point (i.e., the initial approximation) in the closed interval $[a, b]$, and let $\{x_n\}$ be the corresponding iteration sequence. Then the point x_n is contained in the closed interval $[a_{n-1}, b_{n-1}]$ for all $n \geq 2$, and hence $x_n \to x_0$. As we already know, this implies that x_0 is the root of the equation (9.1). It is the only root of this equation. Indeed, if $f(x_0) = x_0$ and $f(y_0) = y_0$, then the contraction condition implies

$$|x_0 - y_0| = |f(x_0) - f(y_0)| \leq \alpha|x_0 - y_0|.$$

Because $\alpha < 1$, we have $|x_0 - y_0| = 0$ which means that the points x_0 and y_0 coincide.

Thus, if f is a contraction mapping of the closed interval $[a, b]$ into itself, then the iteration sequence converges to the unique root of the equation (9.1).

An error estimate for the approximate value x_n of the root can be obtained similarly to the derivation of inequality (9.3):

$$|x_n - x_0| \leq \alpha^n|x_1 - x_0|.$$

That is, the error decreases with the increase of n at least in a geometric progression with the ratio α.

For more details on the iteration method the reader can consult the book [4] and the article [2].

Note that the condition for the convergence of an iteration process yields the following fixed point theorem: *any contraction mapping of the real line into itself has a unique fixed point* (see Problem 32).

Problems. 29. Construct an iteration sequence (and draw the corresponding iteration paths) for the function $f(x) = 1/x$ in the closed interval $[1/2, 2]$ taking for the initial approximation $x_1 = 1$, and then $x_1 = 2$.

30. Suppose that the function $f(x)$ is increasing and continuous in the closed interval $[a, b]$, and that $f(a) < 0$, $f(b) > 0$. Construct the iteration paths taking for the initial approximation $x_1 = a$, and then $x_1 = b$.

Prove that both iteration processes converge (although f is not necessarily a contraction). Do the roots obtained in both processes coincide?

31. Prove a more specified version of the intermediate value theorem. Let $f(x)$ be a continuous function defined on the closed interval $[a, b]$. Let A be the least and B the greatest value of $f(x)$ in this closed interval. This means that $f(x) \geq A$ for all x in the interval, and that there is a point x_1 in it such that $f(x_1) = A$. Similarly, $f(x) \leq B$ for all x and there is a point x_2 such that $f(x_2) = B$. Let $A < c < B$. Then there is a point x_0 in the closed interval $[a, b]$ such that $f(x_0) = c$.

32. Let f be a contraction mapping of the real line into itself. Prove that there is a closed interval on this line mapped into itself by f.

33. Can the iteration approach be applied to the equation $x = 1/(4 + x^2)$? Is there a closed interval that the mapping $f(x) = 1/(4 + x^2)$ takes into itself?

10

Retraction

The fixed point property is closely related to the notion of retraction. Let X be a set and Y its subset. The subset Y is said to be a *retract* of set X if there exists a continuous mapping $f : X \to Y$ preserving each point of the set Y, i.e., identical on Y. The mapping f is then called a retraction of the set X onto Y.

Thus, a closed interval is a retract of a square (Example 1), as well as of a triangle, a convex polygon, a cube, etc. A circle is a retract of an annulus, the region bounded by a pair of concentric circles.

The following theorem explains the importance of this notion.

THEOREM (on retraction). *If a set X has the fixed point property, and Y is a retract of X, then Y also has the fixed point property.*

PROOF. Let $g : X \to Y$ be a retraction, and let $f : Y \to Y$ be any continuous mapping. Consider the composition $\psi = fg$ of these mappings (the mapping g acts first, and f acts second). Then ψ is a continuous mapping of set X into Y, and because Y is contained in X, the mapping ψ can be considered a continuous mapping of X into itself. Hence ψ has a fixed point x_0, which, evidently belongs to set Y. Since the mapping g is identical on Y, the mappings ψ and f coincide on Y. Thus, x_0 is a fixed point of the mapping $f : Y \to Y$.

Because a square has the fixed point property, and a closed interval is a retract of a square, the theorem on retraction provides a new proof of the fixed point theorem for a closed interval. The same argument shows that all other retracts of a square (circle, triangle, or "tailed" square) also have the fixed point property (see Figure 29).

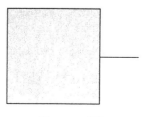

FIGURE 29

One can easily see that a circle does not have the fixed point property. A rotation by the angle equal, say, to $\pi/2$ displaces every point of the circle. On the other hand, a disc has the fixed point property, whereby we obtain an important corollary: *there is no retraction of a disc onto its boundary circle* (see Example 3).

Thus, the statement that no disc can be retracted onto its boundary is a consequence of the fixed point theorem.

Conversely, the fixed point theorem is an easy consequence of the theorem that there is no retraction of a disc onto its boundary.

We give the proof by *reductio ad absurdum*. Suppose that there is a continuous mapping $g : K \to K$ of the disc K into itself that has no fixed points. That is, let $g(p) \neq p$ for all points p in K. For each p consider the ray connecting p with $g(p)$ and directed from $g(p)$ to p. Let $f(p)$ be the point where the ray intersects the boundary circle C (see Figure 30).

Then, associating to each point p the corresponding point $f(p)$, one obtains the mapping $f : K \to C$ such that all points of the boundary circle C are its fixed points. This is a contradiction provided we can show that f is a continuous mapping.

Take any $\varepsilon > 0$ and consider the ε-neighborhood of the point $f(p)$ on the circle C (this neighborhood is evidently an arc). Since the mapping $g : K \to K$ is continuous at point p, for any $\alpha > 0$ there is $\delta > 0$ such that each point from the δ-neighborhood of point p is mapped into the α-neighborhood of point $g(p)$. Take α so small that any ray issuing from any point of the α-neighborhood and passing through any point of the δ-neighborhood (which depends on α) has a common point with the circle C inside the ε-neighborhood of the point $f(p)$ (see Figure 30). This implies the continuity of the mapping f.

Thus, we have deduced the fixed point theorem from the theorem stating that there is no retraction of a disc onto its boundary. To complete the proof we must provide an independent proof of the latter. However, we cannot dwell on this question here.

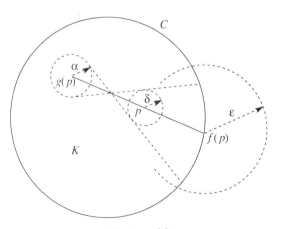

FIGURE 30

To apply the notion of retraction let us come back to the problem from mechanics at the end of Chapter 7. Suppose that the train can now move along any plane curve, and the ordinary hinge connecting the rod to the floor is replaced a spherical one so that the rod can now tilt and fall to any side. As before, we can prove that there is an initial position of the rod ensuring that the rod never falls throughout the journey.

Now the position of the rod is determined by two coordinates: the angle φ measuring its tilt from the vertical position, and the angle α measuring the deviation of its horizontal projection from a fixed direction on the car's floor. It is convenient to interpret the pair of numbers (φ, α) as the polar coordinates of a point on a plane: φ is its radius vector, and α is the polar angle. One can easily see that they satisfy the following conditions: $0 \leq \varphi \leq \pi/2$, $0 \leq \alpha \leq 2\pi$. In other words, the position of the rod at any moment of time is determined only by a point on the disc of radius $\pi/2$. This disc is said to be the phase space of our system.

Suppose that whatever its initial position, the rod always falls. Its coordinates after the fall are of the form $(\pi/2, \alpha)$. Hence, by associating with each initial position of the rod its corresponding final position, one defines a continuous mapping of the disc onto its boundary, which, furthermore, is identical on the boundary. The latter follows from the fact that if the rod was lying on the floor it would stay where it was during the whole journey. Because mapping of that kind is impossible, the statement is proved.

Problems. 34. Do the following figures have the fixed point property: (a) an annulus; (b) a figure eight, that is, a set consisting of two circles with a single common point?

35. Is there a retraction of a square onto the following figures: (a) a "triode" (the letter T figure); (b) an "n-ode" (the figure consisting of n segments issuing from the same point and having no other points in common); (c) a figure eight?

36. Is a closed interval a retract of (a) a triode; (b) a circle; (c) a set consisting of two segments with no points in common?

37. Is a set consisting of two distinct points a retract of a closed interval?

38. Is an annulus a retract of a disc?

39. Is a retract of a compact set compact?

40. Does there exist a continuous mapping of a square onto its boundary not necessarily identical on its boundary?

41. Prove the following "intermediate value" theorem for a disc. Let K be a disc in the plane $\mathbf{R}^2$, let C be its boundary circle, and let $f : K \to \mathbf{R}^2$ be a continuous mapping of the disk into the plane identical on the circle C. Then f "assumes all values in K." That is, for any point q in the disc K there is a point p in K such that $f(p) = q$.

42. Let $f : K \to \mathbf{R}^2$ be as in the preceding problem, and let $g : K \to K$ be any continuous mapping of the disc K into itself. Prove that there is a point p in the disc K such that $f(p) = g(p)$. Does the same statement hold if g is a mapping of the disc K into the plane $\mathbf{R}^2$?

11

Continuous Mappings of a Circle, Homotopy, and Degree of a Mapping

As we know, a circle does not possess a fixed point property. Some of its continuous mappings into itself (for example, the identical mapping) have fixed points, while other mappings (for example, a rotation by a small but nonzero angle) have not. To clarify the situation with these mappings we now present a classification introduced by Brouwer. Two continuous mappings f and g of the circle C into itself are said to be *homotopic* to one another if one mapping can be transformed or deformed into the other continuously. More precisely, there is a family $f_t : C \to C$ of continuous mappings of the circle C into itself depending continuously on the parameter t, $0 \leq t \leq 1$ such that $f_0 = f$ and $f_1 = g$. This family is called a *homotopy* of f to g, and the mappings f and g themselves are said to be *homotopic*.

For example, consider the two mappings in Figure 31. One mapping is the identical one; the other has three foils. (The mappings should be understood as follows: the inner circle is the image, and the outer curve the inverse image. The mapping is defined by projecting the outer curve on the inner one along the radii.)

The foils of the outside curve can be removed by a continuous deformation (see the consecutive positions of the transformation in Figure 32 on p. 48). Then the second mapping is transformed into the first continuously. Hence the two mappings belong to the same class and are therefore homotopic. On the other hand, the mapping shown in Figure 33 on p. 48, which winds the circle twice around itself, can not be continuously deformed into the identical one. Therefore, the two mappings belong to different classes and

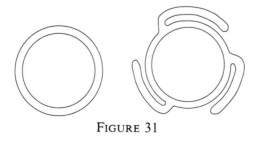

FIGURE 31

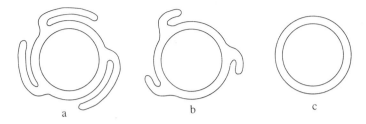

FIGURE 32

FIGURE 33

are not homotopic. As we explained in Chapter 7, the position of a point x on the circle is determined by its angular coordinate α. We assumed that α takes values between 0 and 2π. We now extend the set of possible values of α. Suppose that a point makes a full journey around the circle from $\alpha = 0$ to $\alpha = 2\pi$ in the positive direction and then continues to move. It is natural to assume its angular coordinate will become greater than 2π. After the second full circle, the angle α assumes the value 4π, then 6π, etc. Similarly, moving in the opposite direction yields negative values for the angular coordinate. Thus, each point of the circle is associated with the infinitely many values of the angle α, each value differing by an integer multiple of 2π.

Let f be a continuous mapping of the circle C into itself, sending the point x with the angular coordinate α into the point $y = f(x)$ with the angular coordinate β. Clearly, β is a function of α: $\beta = \varphi(\alpha)$. We can assume that is defined on the closed interval $[0, 2\pi]$. In other words, of all the values of the angular coordinate α, we chose those belonging to the closed interval $0 \le \alpha \le 2\pi$. Since the angular coordinate β is multivalued, the function $\varphi(\alpha)$ can also take many values. It is defined up to $2k\pi$, where k is an integer. Figure 34 shows the graph of the multivalued function corresponding to the identical mapping $y = x$ (for convenience, the scales along the α and β axes are chosen differently). The mapping is seen to have many "branches."

Suppose we chose some value of the function $\varphi(\alpha)$ at the point $\alpha_0 = 0$ in the closed interval $[0, 2\pi]$ such that $0 \le \varphi(0) < 2\pi$ (we will stick to this choice from now on). Let α_1 be another point of the closed interval $[0, 2\pi]$. When the point x moves along the circle so that its angular coordinate changes continuously from 0 to α, the angular coordinate $\beta = \varphi(\alpha)$ of

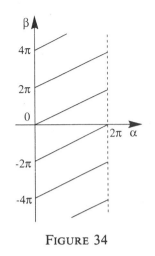

FIGURE 34

the point $y = f(x)$ also undergoes a continuous change and for $\alpha = \alpha_1$ takes a uniquely defined value $\beta = \varphi(\alpha_1)$. This defines a singlevalued continuous branch of the multivalued function φ on the closed interval $[0, 2\pi]$. We will consider this branch in reference to the function φ.

In particular, if α_1 takes the value 2π, we get the equality

$$(11.1) \qquad\qquad \varphi(2\pi) - \varphi(0) = 2n\pi,$$

where n is an integer, which may be positive, negative, or zero. Equality (11.1) follows from the fact $\varphi(0)$ and $\varphi(2\pi)$ are angular coordinates of the same point of the circle.

Equality (11.1) has the following geometric meaning. When the point x makes a single trip around the circle in the positive direction, the image point $f(x)$ also moves along the circle continuously, but not necessarily uniformly. It may speed up, slow down, even stop for some time, or go backwards. But, eventually, after completing several turns (equal to the integer n), it will return to its original position. It is precisely this property that is expressed by equality (11.1).

The number n is called the *degree* of the mapping f and is denoted $\deg f$.

Thus, to any continuous mapping $y = f(x)$ of the circle C into itself there corresponds a unique continuous function $\beta = \varphi(\alpha)$ defined on the closed interval $[0, 2\pi]$ and satisfying condition (11.1) at its end points. Conversely, any such function φ defines a continuous mapping of the circle into itself. For example, Figure 35 on p. 50 shows the graph of the function φ corresponding to the second mapping in Figure 31 on p. 47.

Now we will derive some properties of the mapping f starting from the properties of the function φ.

Any two continuous functions defined on a closed interval, or, in other words, any two continuous mappings $\varphi_0(\alpha)$ and $\varphi_1(\alpha)$ of the closed interval into the real line, are homotopic. The homotopy connecting them can be

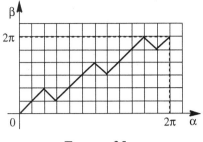

FIGURE 35

defined by the formula

(11.2) $\varphi_t(\alpha) = (1-t)\varphi_0(\alpha) + t\varphi_1(\alpha), \quad 0 \le t \le 1.$

Here the parameter t can be interpreted as the time elapsed after the deformation started, and the deformation itself can be interpreted as a uniform vertical motion that shifts, for each α, each point of one graph into the corresponding point of the other graph.

Let us now show that two continuous mappings f_0 and f_1 of a circle into itself are homotopic if and only if they are of the same degree.

Indeed, suppose that the mapping f_0 is being continuously deformed into the mapping f_1. Then the same deformation affects the corresponding functions $\varphi_0(\alpha)$ and $\varphi_1(\alpha)$, and the differences $\varphi_0(2\pi) - \varphi_0(0)$ and $\varphi_1(2\pi) - \varphi_1(0)$ are equal because each difference describes the change of the angular coordinate of one and the same point on the circle. Thus, any continuous deformation preserves the degree of mapping.

Conversely, let the mappings f_0 and f_1 have the same degree n. Consider the corresponding functions $\varphi_0(\alpha)$ and $\varphi_1(\alpha)$ and connect them with the homotopy (11.2). Then each "intermediate" function $\varphi_t(\alpha)$ satisfies equality (11.1):

$$\varphi_t(2\pi) - \varphi_t(0) = (1-t)\varphi_0(2\pi) + t\varphi_1(2\pi) - (1-t)\varphi_0(0) - t\varphi_1(0)$$
$$= (1-t)[\varphi_0(2\pi) - \varphi_0(0)] + t[\varphi_1(2\pi) - \varphi_1(0)]$$
$$= (1-t)2n\pi + t \cdot 2n\pi = 2n\pi.$$

This means that the family of functions $\varphi_t(\alpha)$ $(0 \le t \le 1)$ corresponds to the family of continuous mappings f_t of the circle into itself. When the parameter t changes from 0 to 1, the point $f_t(x)$ moves along the circular arc from the point $f_0(x)$ to the point $f_1(x)$ uniformly. In other words, the mappings f_0 and f_1 are homotopic.

Thus, the degree of a mapping is a characteristic property of the whole class of homotopic mappings.

As an example, consider a linear mapping f_n of the circle into itself defined by the formula

(11.3) $\varphi_n(\alpha) = n\alpha, \quad n \ne 0, \quad n$ is an integer.

The mapping has the following "geometric meaning." When the point x corresponding to the angle α makes one full circle in the positive direction, its image $y = f_n(x)$ corresponding to the angle $n\alpha$ makes n full circles in the positive direction if $n > 0$, or in the negative direction if $n < 0$. The graph of the function $\varphi_n(\alpha)$ is the line segment connecting the points with the coordinates $(0, 0)$ and $(2\pi, 2n\pi)$. This implies that the degree of the mapping f_n equals n. Consequently, for different m and n the mappings f_m and f_n are not homotopic.

These results imply that all continuous mappings fall into infinitely many classes, one for each integer n $(n = 0, \pm1, \pm2, \dots)$. The mappings of the same class are homotopic and have the same degree n. A representative of such a class is the linear mapping f_n defined by formula (11.3). In particular, the mapping (11.3) is defined for $n = 0$, and maps the circle into a single point. This provides a complete homotopic classification of mappings described at the beginning of Chapter 11.

Now we prove that any continuous mapping $f : C \to C$ of the circle into itself of the degree n has at least $|n - 1|$ fixed points. Consider the function $F(\alpha) = \varphi(\alpha) - \alpha$ defined on the closed interval $[0, 2\pi]$, where $\varphi(\alpha)$ is a single-valued branch of the angular coordinate of point $f(x)$. Equality (11.1) implies that

$$F(2\pi) - F(0) = 2(n - 1)\pi.$$

Hence, if $F(0) = \varphi(0) = \beta_0$ then $F(2\pi) = \beta_0 + 2(n - 1)\pi$. The closed interval with the end points β_0 and $\beta_0 + 2(n-1)\pi$ has the length $2\pi|n - 1|$ and consequently contains at least $|n - 1|$ numbers equal to consecutive integer multiples of 2π. By the intermediate value theorem, there are such α_i $(i = 1, 2, \dots, |n - 1|)$ on the closed interval $[0, 2\pi]$ for which the function $F(\alpha)$ assumes the values equal to the above numbers. That is,

$$F(\alpha_i) = 2\pi k_i, \quad k_i \text{ is an integer.}$$

Hence

$$\varphi(\alpha_i) = \alpha_i + 2\pi k_i, \quad i = 1, 2, \dots, |n - 1|.$$

The last equality shows that the points of the circle C with the angular coordinates α_i are fixed points of the mapping $f : C \to C$.

The case where $\deg f = 1$ is special. There are mappings of degree 1 that have fixed points, while other mappings have none. (Note that degrees of both mappings given at the beginning of Chapter 11 equal 1.)

Problems. 43. Prove that any two continuous mappings f_0 and f_1 of a closed interval (as well as those of a circle or a square) into the real line are homotopic.

44. Consider an ordinary clock with two hands, one showing hours, the other minutes. At each moment of time the position α of the hour hand (measured in hours ranging between 0 and 12, or in degrees changing between 0 and 360, or in radians between 0 and 2π) corresponds to the unique position $f(\alpha)$ of the minute hand. For example, if $\alpha = 3\frac{1}{2}$ hours then $f(\alpha) = 6$ hours; if $\alpha = 5$ hours, then $f(\alpha) = 0 = 12$ hours. This defines a continuous mapping f of the circle into itself. Find the degree and fixed points of this mapping. Is it possible to define the mapping φ of

the circle into itself by associating to each position of the minute hand the position of the hour hand at the same moment of time?

45. Consider a "broken" clock in which (a) starting from 0 hours the minute hand moves twice as rapidly as the hour hand; (b) starting from 0 hours the minute hand moves with the same speed as the hour hand but in the opposite direction; (c) at each moment of time the positions of the minute and hour hands differ by two hours. The mapping f is defined as in Problem 44. Find the degree of the mapping and its fixed points in all three cases.

12

Second Definition of the Degree of a Mapping

Theories of continuous mappings and fixed points for a circle and for a sphere are similar in many respects. In particular, the notion of the degree of a mapping also plays an important role for a sphere. However, the approach used in the last chapter to define the degree in the case of a circle is difficult to apply for a sphere. That is why we give an equivalent definition of the degree of mapping of a circle into itself that admits a straightforward generalization to the case of a sphere.

The main idea is to count the number of inverse images for each point of the circle. It is based on the observation that in some (algebraic) sense this number is the same for all points.

For example, consider once again the linear mapping $\beta = \varphi_n(\alpha) = n\alpha$ for $n \neq 0$. It sends the points of the circle whose angular coordinates differ by $k/|n| \cdot 2\pi$ ($k = 0, 1, \ldots, |n| - 1$) into one and the same point. In other words, each point of the circle C has exactly $|n|$ inverse images.

Now let f be a *piecewise linear mapping* of the circle into itself, i.e., a mapping the graph of which is a broken line (see Figure 36, where $\deg f = 3$).

In order to find the inverse images of a point on the circle with the angular coordinate β_0 ($0 \leq \beta_0 < 2\pi$) let us draw the horizontal lines $\beta = \beta_0 + 2k\pi$ ($k = 0, 1, \ldots, |n| - 1$) and find the points where these lines intersect the

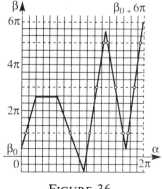

FIGURE 36

53

graph of $\beta = \varphi(\alpha)$. To each intersection point we associate a sign: plus if the function $\varphi(\alpha)$ is increasing at this point and minus if it is decreasing. Counting the number of inverse images, for example, for the angle $\beta_0 = \pi$, we get $7 - 4 = 3$, i.e., exactly the number required by the degree of the mapping.

Note that we did not count the points at which the graph of the function $\varphi(\alpha)$ was tangent to horizontal lines (i.e., the vertices of the angles on the graph), or the points of the horizontal sections of the graph itself. We could as well have ascribed to each such point both signs (plus and minus) simultaneously, which clearly would not have affected the algebraic sum of the number of inverse images.

In general, for any piecewise linear mapping of the circle into itself the number of inverse images (taken with appropriate signs) is the same for all points. Indeed, we have already seen that this statement holds for the linear mapping $\varphi_n(\alpha) = n\alpha$, $n \neq 0$. Now any piecewise linear mapping can be obtained from the linear mapping by a continuous deformation (homotopy). Under the deformation, for some of the points new inverse images can arise, or old inverse images disappear, but they can come and go only in pairs, one with the plus sign, and another with the minus. Hence, the number of inverse images for any point is always the same, and equals precisely the degree of mapping in the sense of the first definition.

Thus we arrive at the second definition: *the degree of a piecewise linear mapping of a circle into itself* is the number of inverse images of each point taken with appropiate signs.

If we now consider an arbitrary continuous mapping f of a circle into itself, then some of the points can have infinitely many inverse images, so that this definition does not work. In such a case, the mapping f is usually replaced by a close piecewise linear mapping g for which $\deg g$ is well defined. One can prove that all piecewise linear mappings g sufficiently close to f have the same degree, which can then be taken for the degree of f.

Problem. 46. A square is triangulated in such a way that all vertices of the triangulation are labeled by the numbers 0, 1, or 2 (one number for each vertex). A separate triangle (which is not triangulated) has its three vertices labeled by the same numbers. One can define the following a piecewise linear mapping of the boundary of the square into the boundary of the triangle. If an edge of the triangulation lying on the boundary of the square is labeled by 0 and 1, then it is mapped linearly onto the side of the triangle bearing the same labels. If both sides of the edge are labeled by 0, it is mapped into the vertex of the triangle having the same label; and similarly for all other combinations of labels. Prove the following generalization of Sperner's Lemma: if the degree of this mapping is different from zero, then there is a face in the triangulation of the square bearing three different labels.

The square can be replaced with any convex polygon.

13

Continuous Mappings of a Sphere

Now we can study mappings of a sphere S. A *great circle* on S is the intersection of the sphere with a plane passing through its center. A *triangulation* of a sphere is its decomposition into spherical triangles (faces) satisfying the usual requirement on their relative position (see the beginning of Chapter 4). The sides of these triangles are always arcs of great circles. Figure 37 shows two triangulations of a sphere obtained by the central projection of the faces of the inscribed octahedron (eight faces) and icosahedron (twenty faces).

Sometimes it is convenient to visualize the sphere topologically as the surface of a convex polyhedron, for example, the octahedron (see Figure 38 on p. 56).

The sphere S and this surface are, of course, only homeomorphic, but we shall consider them "identical." The faces of the octahedron will be triangulated into smaller triangles. We will maintain this point of view while considering piecewise linear mappings.

Now consider two triangulations of the sphere S, in general, different from one another. A continuous mapping of the sphere into itself is said to be *piecewise linear* if each triangular face of the first triangulation is mapped linearly[2] onto a triangular face, edge, or vertex of the second triangulation.

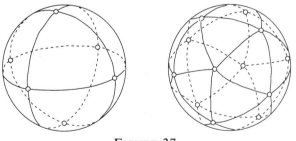

FIGURE 37

---

[2] A linear mapping of one triangle onto the other can be defined by two linear functions in two variables, or as the mapping preserving barycentric coordinates (on the latter see the book [1]).

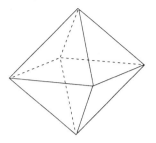

FIGURE 38

The sphere is *oriented* as follows. Let a point move over its surface along a small circle (or along the boundary of a triangular face), and position an observer outside the sphere near the circle. Then the point is said to move in the positive direction if the observer sees it moving counter-clockwise.

Let ABC be a face of the triangulation of the sphere, and suppose that the mapping f maps it into the face $A_1B_1C_1$ of the other triangulation so that $f(A) = A_1$, $f(B) = B_1$, and $f(C) = C_1$. Let the sequence $ABCA$ define the positive orientation on the boundary of the face ABC. The face is said to be *positive* if its orientation is preserved by the mapping f, i.e., if the sequence of points $A_1B_1C_1A_1$ defines the positive orientation of the boundary of $A_1B_1C_1$. A face ABC is said to be *negative* if the mapping reverses the orientation.

Consider an arbitrary piecewise linear mapping of one triangulation of the sphere into another. Take any face of the second triangulation and count the difference between the numbers of positive and negative faces of the first triangulation mapped onto it (i.e., the number of the inverse images of the face taken with the corresponding signs). It turns out that *these differences are the same for all faces of the second triangulation.* Their common value is called the *degree* of the piecewise linear mapping.

We shall prove this statement (Theorem on the Degree) by our "walking process." Take any two adjacent faces ABC and ADB of the second triangulation (see Figure 39, where they are viewed from outside of the sphere). We shall now prove that the numbers of inverse images for them taken with their signs are equal.

The definition of a piecewise linear mapping immediately implies the following statement. If F_1 is a face of the first triangulation mapped onto ABC

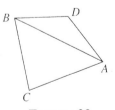

FIGURE 39

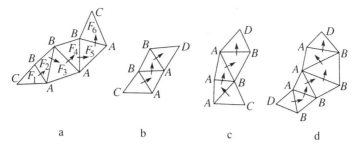

FIGURE 40

then there is a chain of faces $F_1, F_2, \ldots, F_{k-1}, F_k$ in this triangulation that has the following properties:

(1) Any two subsequent faces in the chain have a common edge;

(2) the last face F_k is mapped either on ABC, or on ADB;

(3) each of the intermediate faces $F_2, \ldots, F_{k-1}$ (which, incidentally, are not necessarily bound to exist) is mapped onto the edge AB.

An example of such a chain is shown in Figure 40a. The letters A, B, and C marking the vertices of the faces in this figure refer to their images. In other words, each vertex marked by A is mapped into the vertex A in Figure 39, and similarly for the other letters. If the face F_1 is mapped on ABC preserving the orientation (that is, if F_1 is positive), then the face F_k, when mapped on ABC, reverses the orientation (face F_k is negative). We can clearly see this from Figure 40a. We shall, therefore, say that the chain of faces from F_1 to F_k defines a path joining $+ABC$ with $-ABC$. By constructing such a chain starting with each face of the first triangulation mapped into ABC or ADB, we obtain paths of the following four types:

$$\text{from } +ABC \text{ to } -ABC,$$
$$\text{from } +ABC \text{ to } +ADB,$$
$$\text{from } -ABC \text{ to } -ADB,$$
$$\text{from } +ADB \text{ to } -ADB.$$

(See Figure 40.) Denote by m, n, p, and q the number of paths of the first, second, third, and fourth types, respectively. Then the number of inverse images of the face ABC taken with their signs equals $m - m + n - p = n - p$ and for the face ADB this number is $n - p + q - q = n - p$. We see that these numbers are equal. The Theorem on the Degree is thereby proved for a pair of faces of the second triangulation having a common edge. The statement can then be verified for the pair of faces ADB and AED with the common edge AD, etc.

In order to define the degree of an arbitrary continuous mapping of the sphere $f : S \to S$, we must, as in the case of the circle, replace it with a close piecewise linear mapping g, find $\deg g$ and verify that any two sufficiently close approximations g_1 and g_2 yield the same value of degree, which is then taken, by definition, to be the *degree* of the mapping f.

A simple example (which we will use immediately) is that the *identical* mapping has the degree 1. On the other hand, consider the *antipodal* mapping sending each point x into its antipode x^*. This mapping reverses the orientation of the sphere (see Figure 41), and has the degree -1. On the circle, however, the antipodal mapping preserves the orientation, and has the degree 1.

LEMMA 5. *Let f and g be continuous mappings of a sphere into itself such that for any point x on the sphere, its images $f(x)$ and $g(x)$ are not antipodes: $f(x) \neq [g(x)]^*$. Then f and g are homotopic and have the same degree: $\deg f = \deg g$.*

PROOF. Indeed, because the points $f(x)$ and $g(x)$ are not antipodal, they can be joined by the unique arc of a great circle whose length is less than that of a semicircle. Hence, by moving the point $f(x)$ (uniformly with respect to all x) into the point $g(x)$ along this arc one obtains a homotopy of f to g. Under a continuous deformation the degree of the mapping changes continuously. Since the degree of the mapping is an integer, it does not change at all.

Of course, Lemma 5 holds for a circle as well.

FIXED POINT THEOREM FOR A SPHERE. *Any continuous mapping f of a sphere S either has a fixed point or sends some point into its antipode. If $\deg g \neq -1$ then g always has a fixed point. If $\deg f \neq 1$ then there is a point on the sphere mapped into its antipode.*

PROOF. Let us first assume that $\deg f \neq -1$, but the mapping has no fixed points. Then one can apply Lemma 5 to the mapping f and the antipodal mapping $g(x) = x^*$. Since $\deg g = -1$, it yields $\deg f = -1$, a contradiction.

Let $\deg f \neq 1$ and $f(x) \neq x^*$ for all points x of the sphere S. Then we apply Lemma 5 to the mapping f and the identical mapping $g(x) = x$ for which $\deg g = 1$. Then $\deg f = 1$, again a contradiction. Suppose that we have no prior information on the degree of the mapping f. If it has no fixed points, then we can conclude, as before, that $\deg f = -1$. Assuming that no point x is mapped by f into its antipode x^*, we can again apply Lemma 5 to the mapping f and the identical mapping. The lemma yields $\deg f = 1$, which contradicts $\deg f = -1$. The theorem is proved.

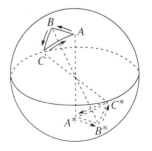

FIGURE 41

Problems. In Problems 47 to 54 a mapping of a circle or a sphere into itself is always assumed to be a *continuous* mapping.

47. Prove that if f is a mapping of a circle into itself, and $\deg f \neq 1$, then f has a fixed point, and, moreover, maps some point of the circle into its antipode.

48. Let f be as in Problem 47. Prove the following two statements: (a) if no point of the circle is mapped into its antipode, then $\deg f = 1$; (b) if the circle is not covered by its image then f has a fixed point. What is the value of $\deg f$ in the last case?

49. A mapping f of a circle into itself is said to be *odd* if it sends any pair of antipodes into a pair of antipodes, that is, if $f(x^*) = [f(x)]^*$ for all points x of the circle. Prove that the degree of an odd mapping is odd.

50. Prove that if f is an odd mapping of the circle into itself, and $\deg f \neq 1$, then there is a pair of fixed antipodal points.

51. For each integer n give an example of a mapping of a sphere into itself having the degree n.

52. Verify that if f_1 and f_2 are mappings of a circle or a sphere into itself and $f_3 = f_2 f_1$ is the composition of the mappings f_1 and f_2 (see p. 18) then

$$\deg f_3 = \deg f_1 \cdot \deg f_2.$$

53. If f and g are mappings of a sphere into itself, then at least one of the three mappings f, g, and gf has a fixed point. In particular, the composition f^2 of any mapping f with itself has a fixed point. Does the same statement hold for a circle?

54. Prove that any mapping of a sphere into itself either has a fixed point or has a pair of points that exchange their positions. Does this statement hold for a circle?

In conclusion, we give a brief overview of vector fields and their singular (or critical) points. Can one construct a continuous *field of directions* on a circle or a sphere. That is, specify a vector at each point in such a way that it changes continuously when one moves from one point to another? The points where the vectors either vanish or are not defined at all are called *singular* points of the field. For example, the north-to-south direction on a sphere (see Figure 42b) has its singular points at the poles, where the vectors look in different directions, thus breaking the continuity of the field. The same is true for the west-to-east direction (see Figure 42c).

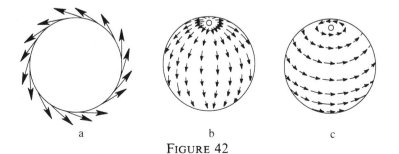

a   b   c

FIGURE 42

FIGURE 43

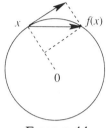

FIGURE 44

A continuous mapping f of a circle into itself naturally defines a vector field, if one considers for each point x the vector joining it with its image $f(x)$. For example, Figure 43 shows the vector field corresponding to the mapping $\beta = 2\alpha$. Since the point $\alpha = 0$ is a fixed point of the mapping, it is also a singular point of the vector field.

If no vector of the field is perpendicular to the circle (in other words, if no point of the circle is mapped by f into its antipode), we can consider the projection of each vector onto the tangent to the circle at the point the vector issues from (see Figure 44), thus yielding the field of tangent directions on the circle. Vectors perpendicular to the circle are projected into zero vectors, so they define singular points of the tangent field. In Figure 43 such a point is $\alpha = \pi$.

As we know (see Problem 47), any continuous mapping of a circle into itself of the degree different from 1 either has a fixed point, or the antipodal property, that is, the mapping takes some point into its antipode. On the other hand, there are mappings of degree 1 that have neither fixed points nor antipodal property. Thus we can conclude that *there exist continuous fields of tangent vectors on a circle; they are generated by mapping of degree 1.*

On a sphere, any continuous mapping has either a fixed point or an antipodal property. Therefore, *there can be no continuous field of tangent directions on a sphere.* This theorem is also known as the Theorem on a Hedgehog. A hedgehog rolled into a ball has at least one needle sticking perpendicular to its body surface.

14

Theorem on Equality of Degrees

In this chapter Sperner's lemma is reformulated as a statement on the degree of a piecewise linear mapping of one polygon into another. In this form it admits a generalization that includes, as a special case, the theorem stating that no polygon can be retracted onto its boundary.

Let T_1 be a triangle with a triangulation whose vertices are labeled by the numbers 1, 2, and 3, as in Sperner's lemma, and let T_2 be a second triangle without triangulation, with vertices also labeled by 1, 2, and 3. Consider these labels from the following point of view. Suppose all vertices of the triangulation labeled by 1 are mapped into the vertex of T_2 labeled by the same number and that the same is true for vertices labeled by 2 and 3. Then all vertices of the triangulation of T_1 are mapped into the vertices of the triangle T_2. Extending this mapping piecewise linearly to all edges and faces of the triangulation (there is only one way to do it), we get the mapping $f: T_1 \rightarrow T_2$. In particular, if a face bears the labels $(1, 1, 1)$ it is mapped into the vertex 1 of the triangle T_2, the face labeled by $(2, 2, 3)$ is mapped onto the side $(2, 3)$. All faces mapped into either vertices or sides of triangle T_2 are said to be degenerate. All nondegenerate faces are of only one type, namely the faces labeled by $(1, 2, 3)$; they are mapped on the entire triangle T_2.

It is now clear that Sperner's lemma can be formulated as follows. Let T_1 and T_2 be two triangles, the first one being triangulated. Suppose that a side of T_2 is assigned to each edge of the triangulation of T_1 and that $f : T_1 \rightarrow T_2$ is a piecewise linear mapping under which each edge of the triangulation of T_1 is mapped onto the corresponding side of the triangle T_2. Then the triangulation has a nondegenerate face.

Let us discuss now a more general problem. Consider a triangulated polygon P that is either convex, or, in a more general situation, has the boundary ∂P that is a closed broken line without self-intersections (in other words, ∂P is homeomorphic to a circle). Consider a pair of such polygons P_1 and P_2 and a piecewise linear mapping $f : P_1 \rightarrow P_2$. This mapping can also be described by assigning numbers to vertices of triangulations of P_1 and P_2 as follows. For P_2 the only requirement for labels is that they distinguish vertices of the triangulation (that is, any two vertices have different labels). In the polygon P_1 the labels of the vertices indicate their images under the

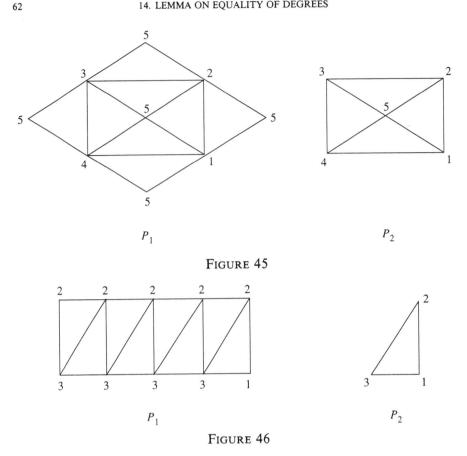

FIGURE 45

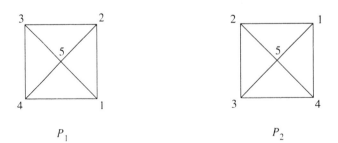

FIGURE 46

mapping f: a vertex of the triangulation of P_1 and its image in P_2 must have the same label. Examples of piecewise linear mappings of polygons follow. Figure 45 depicts the mapping "making an envelope out of a sheet of paper."

In Figure 46 all but the extreme right triangular faces of the rectangle are mapped onto the edge (2, 3) while the rectangle goes into the triangle.

Figure 47 shows the rotation of the square by $\pi/2$.

To formulate the main theorem of this chapter we need to introduce the notion of the degree of a piecewise linear mapping. Before we give the defi-

FIGURE 47

nition, let us recall how a plane polygon can be oriented. To orient a polygon is to choose one of the two directions on its boundary and call it the positive direction. Usually, a positive direction is chosen counter-clockwise. In other words, if you follow the boundary in the positive direction, the interior points of the polygon stay on your left. The reverse direction is said to be negative. Each face of the triangulation is oriented accordingly, so that its orientation agrees with the orientation of the triangle itself. Therefore each inner edge of the triangulation acquires two opposite orientations inherited from two adjacent faces to which it belongs.

Let P_1 and P_2 be two triangulated and oriented polygons, and $f : P_1 \to P_2$ a piecewise linear mapping that maps the boundary ∂P_1 into the boundary ∂P_2. Now, as for the sphere, the degree of the mapping $f : P_1 \to P_2$ can be defined as the difference in the number of positive and negative faces of polygon P_1 mapped into a fixed (and therefore arbitrary) face of polygon P_2. Since the boundary ∂P_1 goes into the boundary ∂P_2, the degree of the mapping $f : \partial P_1 \to \partial P_2$ is also well defined. The first degree will be called the *inner* degree and the second the *boundary* degree.

THEOREM ON EQUALITY OF DEGREES. *For any piecewise linear mapping of two polygons the boundary and the inner degrees are equal.*

The proof of the theorem is left to the reader, who can obtain it, for example, by walking the faces of the polygon P_1.

We can easily see that in Sperner's lemma the boundary degree equals 1, so that the inner degree is also 1. This means that at least one of the faces of the triangulation is mapped onto the entire triangle T_2. Thus, the theorem on the equality of degrees is a generalization of Sperner's lemma.

Suppose now that the polygons P_1 and P_2 coincide and have the same triangulation, and suppose that f maps the entire polygon P $(= P_1 = P_2)$ into its boundary. Since all triangular faces become degenerate, the inner degree is 0. If the mapping f were to map the boundary ∂P onto itself identically, the boundary degree would equal 1. This would contradict the theorem on the equality of degrees, and we get the following corollary. There is no piecewise linear retraction of a polygon onto its boundary. As we already know, this statement holds for any continuous (and not only piecewise linear) retraction.

Problems. 55. Deduce Lemma 1 and its refined version from the theorem on the equality of degrees.

56a. Prove there is no piecewise linear mapping of a triangulated square onto its boundary such that each point of the boundary goes into its antipode (that is, the opposite point of the boundary).

56b. Prove that there is no piecewise linear mapping of a triangulated square onto its boundary such that each pair of antipodes at the boundary goes into the pair of antipodes. This problem is related to Problems 6 and 49.

57. Does the theorem on the equality of degrees hold without the assumption that the boundary of the polygon is mapped into the boundary?

Solutions and Answers

1. The functions $f_2(x)$, $f_3(x)$, $f_4(x)$.

4. Let f be a nontrivial involution of the closed interval $[0, 1]$. Then the reflection in the line $y = x$ takes its graph into itself. Therefore, if the graph has two distinct points on this line, they are joined by two different curves contained in the graph. This is impossible because the function f is single-valued in all points of $[0, 1]$.

5. Rooms are closed subintervals. For doors we can take the left endpoint of the original interval and the points of the subdivision labeled by 0. Dead ends are closed subintervals bearing different labels.

6. Suppose that there are no edges of the type $(1, -1)$ and $(2, -2)$. Then there are evidently no triangular faces labeled by three different numbers. Let us prove that the boundary of the square Q contains an even number of edges of the type $(1, -2)$. We shall call such an edge a door. Dead ends must be triangular faces with three different labels, two of which are 1 and -2, but there are no such faces. Therefore, if we start a walk over the faces of the triangulation by entering the square through an outside door $(1, -2)$, the walk can be terminated only if we exit Q through another outside door $(1, -2)$. This means the number of edges of type $(1, -2)$ on the boundary of the square Q is even.

Now let A be the left bottom vertex of the square Q, let B be the opposite vertex, let S be the part of the boundary of Q consisting of the bottom side and right side, and let S' be the opposite part of the boundary. Each edge of type $(1, -2)$ lying on S' corresponds to the edge of type $(-1, 2)$ on S. Hence the total number of boundary edges of type $(-1, 2)$ is the sum of the number of edges of type $(1, 2)$ on S and the number of edges of type $(-1, 2)$ also lying on S. However, this sum equals the total number of changes in signs of labels encountered when we follow S from the point A to the point B. Because this number is odd, we get a contradiction.

7. This problem is due to H. Steinhaus (see [6, Russian page 19], [7, pages 103 and 211]). Interestingly, the solution given in the book [7] and first published in *Matematica* was later rejected by the author as incorrect.

We will say that a set of castle fields is a *rook array* (or simply an *array*) if a rook can move from any of its fields to any other field of the same set by following only the rook fields. The set of all rook fields of the chessboard

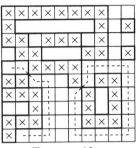

FIGURE 48

then falls into separate arrays so that if a rook is positioned in one of them, no other array is accessible to it. Thus, no pair of arrays has common fields, but some pairs may have common corner points.

An array can have holes inside that either consist of king's fields exclusively or include other arrays. An array with no holes has a single (exterior) boundary. Otherwise it may also have (one or more) interior boundaries.

We take as obvious the statement that the exterior boundary of an array is a closed broken line without self-intersections. Hence, if an array has no fields lying at the edge of the board, the king can circle the array freely by moving over his own fields adjacent to the array's exterior boundary. If an array does include an edge field of the board, the king can never complete a circular journey; thus the journey always ends at the edge of the board. A common corner point of two arrays is not an obstacle for king's walk. This point is illustrated in Figure 48 where the rook's fields are marked with a cross and the king's routes are shown by dotted lines.

Now assume that the statement of the problem is false, so that, for any initial position, the rook cannot cross the board from the bottom edge to the top edge by moving only over its own fields. Let us prove that then the king can find his way from the left edge of the board to the right edge.

Suppose that the rook, after starting its journey from a field of the first (bottom) row, has reached the ith row but is unable to reach the $(i + 1)$th row (see Figure 49).

Hereafter we consider only arrays having fields in the ith row. Some arrays we will call "suitable" and the others "unsuitable." An array is "suitable" if it has at least one field in the first row; otherwise it is "unsuitable." If a

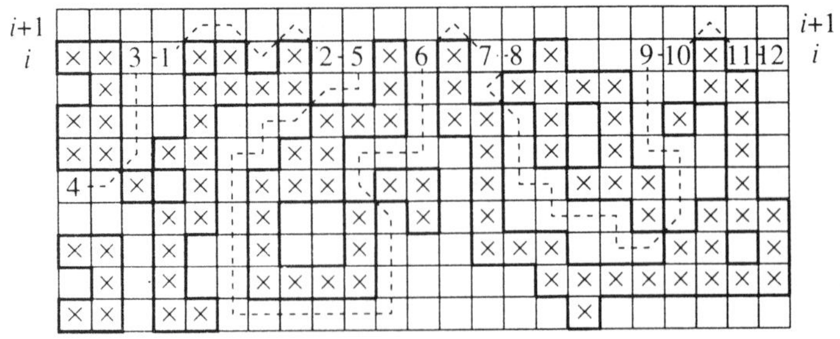

FIGURE 49

rook is in a "suitable" array, it can find its way from the first row to the ith row, and if the array is "unsuitable," the rook is unable to do that. By the assumption, there is at least one suitable array.

By the same hypothesis, there must be a king's field above each field of a suitable array lying in the ith row. Therefore, all suitable arrays are contained in the part of the board formed by the first, second, ..., ith rows. There is no such restriction for unsuitable arrays.

Let us now examine the ith row in more detail. If all the rook's fields in this row belong to the same array (which is then necessarily a suitable one), the king can easily traverse the board from left to right. If the entire ith row consists of rook's fields, then the king can move over the whole length of the $(i + 1)$th row; otherwise the king moves over the king's fields of the ith row and those fields of the $(i + 1)$th row that are positioned above the rook's fields of the ith row.

Thus we can assume there are rook's fields in the ith row belonging to different arrays. Let us walk along the ith row from left to right, registering the following groups of fields in the order we encounter them:

(1) as we start our journey, we encounter either a group of king's fields, which we shall call the extreme left group, or the group of rook's fields belonging to the same array, possibly interspersed with some king's fields;

(2) a group of consecutive king's fields, which we shall call a separating group;

(3) the second group of rook's fields belonging to one array that is different from the first array and again possibly interspersed with king's fields;

(4) another separating group of consecutive king's fields, etc. up to the right edge of the chessboard; the last one is either a group of rook's fields belonging to some array (and possibly interspersed with king's fields), or the extreme right group of king's fields.

Note that some of the groups (and even all of them) may consist of a single field. Another important observation is that different (but not neighboring) groups of rook's fields may belong to the same array, being joined by fields lying outside the ith row.

Section by section we shall plot the king's route from the left edge of the board to the right edge, trying to steer the king through separating and extreme groups of fields. Take a suitable array and a group of the king's fields in the ith row. In general, there are separating groups of fields in the ith row to the left and to the right of him. The king is then conducted from the left separating group to the right one through the fields of the ith and $(i + 1)$th rows (see Figure 49, where the beginning and the end of this section of his journey are labeled by the numbers 1 and 2, respectively). Then the king follows the left separating group from right to left until he runs into the next array (point 1 to point 3 in Figure 49). If this array is suitable, the king can bypass it through the fields of the ith and $(i + 1)$th rows by moving from right to left until he meets another separating (or extreme left) group. If the array if unsuitable, the king starts to circle it clockwise, moving over the fields adjacent to the array's boundary. This array contains no points of the first row, so that the king reaches either another separating (or extreme left) group of fields or the left edge of the board (the route from point 3 to point 4 in Figure 49).

By repeating this procedure the king finally finds its way from the "midpoint" of the ith row to the left edge of the board, i.e., in the direction opposite to the desired one. A similar procedure steers the king from the midpoint to the right edge of the board. The only difference is that now the king moves above suitable arrays from left to right and circles unsuitable ones counter-clockwise. The complete route of the king in Figure 49 looks like this: point 4 to point 3, 3 to 1, 1 to 2, 2 to 5, 5 to 6, and so on until point 12 is reached.

The statement of the problem holds if the king is replaced by a queen, or if the king and the rook change roles. The situation is different for two rooks: assigning white fields to one and black fields to the other makes it impossible for either rook to move at all.

This problem is used in the proof of the fixed point theorem (see Chapter 8).

8. The proof is obtained using a "walking process." Suppose all fields of the board are occupied by white and black pieces in such a way that the two white sides are occupied by white pieces, and the two black sides are occupied by black pieces. The corner fields may be occupied by either black, or white pieces. By joining each pair of neighboring hexagonal fields by line segments we get ("dual") rhombus lattice triangulated into regular triangles (Figure 50a). Each hexagonal field corresponds to the unique vertex of the triangulation. Let us label the vertices corresponding to white fields (pieces) by 0, and label those corresponding to black pieces by 1. We shall say that an edge of the triangulation is a door if its endpoints are labeled differently. It is easy to see that no face is a dead end. Thus, a walk started in one corner must inevitably end in another corner. The route shown in Figure 50b joins the bottom side of the rhombus with the top side, which means the whites won.

This problem, as the preceding one, can be used for proving the fixed point theorem.

10. Let m be the number of all edges, n the number of all vertices, and $k_1, \ldots, k_n$ the multiplicities of these vertices. Then

$$k_1 + k_2 + \cdots + k_n = 2m,$$

because each edge contributes to the multiplicities of its endpoints. Hence, the number of odd terms in the left-hand side of the equality must be even.

11. Let the vertices of each face of the polygon be labeled by the numbers 0, 1, and 2. Put an arrow on each edge directed from 0 to 1, from 1 to 2, and from 2 to 0. Then each face acquires an orientation—some faces clockwise, others counter-clockwise. The adjacent faces are oriented differently. Hence, by coloring the faces of the first type white and those of the second type black, one obtains a regular coloring.

Conversely, suppose the faces of the triangulation admit a regular coloring. If an edge of the triangulation is a side of a black triangle, then we choose

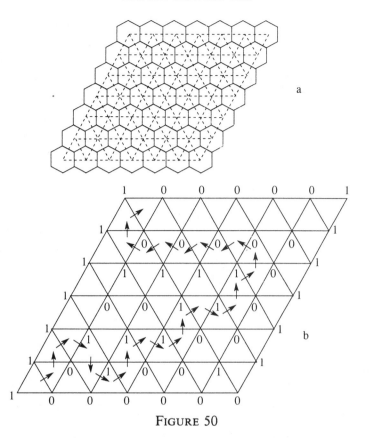

FIGURE 50

the direction on it with respect to which the interior points of this face lie on the left. Then each black face is oriented counter-clockwise, and some of the white ones (all except those lying on the boundary of the polygon) are oriented clockwise (see Figure 51a).

Suppose we have started our walk at some vertex of the triangulation, moved along a closed (possibly self-intersecting) path always in the direction of the arrows, and returned to the same vertex. Then the number of edges we have passed is divisible by 3. Indeed, deleting the faces of the triangulation

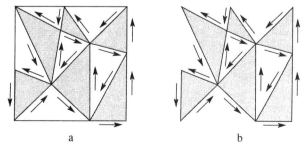

FIGURE 51

lying outside our path, we get a polygon P whose boundary borders the faces of one color only (e.g., black) (see Figure 51b). The total number of boundary edges of the polygon P equals the difference between the total number of edges of its black faces and the total number of edges of all its white faces. Because the last two numbers are divisible by three, their difference is also divisible by three.

Thus, the number of edges in such a closed path is divisible by three. It follows therefore that if a vertex A is joined with a vertex B by two different paths going along the edges, such that there are p edges in the first path and q in the second, then the difference $p - q$ is divisible by 3.

Clearly, any vertex of the triangulation can be reached by such a path from any other vertex, at least if we do not care about the direction of the edges. However, the path can always be modified in such a way that direction of the motion coincides with that of the arrows: namely, each edge passed in the direction opposite to the arrow can be replaced with two other edges of an adjacent face. A similar detour around the adjacent face can replace a section of the path that is a boundary edge of the original polygon and is not equipped with an arrow.

Now take a vertex A, and let B be any other vertex. A path from A to B along the arrows can be plotted in different ways. Different paths may include different numbers of edges, but the previous arguments guarantee that the remainder of all these numbers when divided by 3 is always the same. This remainder is assigned to the vertex B as its label. Repeating the process for all other vertices, one labels all of them by the numbers 0, 1, and 2, and, evidently, three vertices of each face carry three different labels.

We have, therefore, proved that the statements (a) and (b) of the problem are equivalent. One can show that each of them is equivalent to the following statement: (c) multiplicities of all interior (i.e., not lying on the boundary of the polygon) vertices of the triangulation are even.

12. Consider the shaded triangles (Figure 52). The number of sides joining vertices of different colors in each of them is even (i.e., equals 0 or 2) whatever the coloring. The total number of edges with endpoints of different colors in

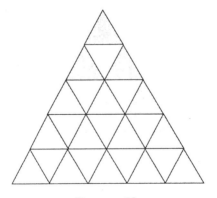

FIGURE 52

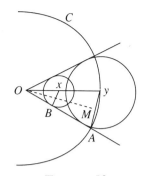

FIGURE 53

the triangulation equals the sum of numbers of such sides for all triangles, and is, therefore, also even.

The statement does not necessarily hold for an arbitrary triangulation of the triangle. For an arbitrary polygon the following statements are equivalent: (a) for any 2-coloring of all vertices of the triangulation the number of edges with endpoints of different colors is even; (b) the multiplicities of all vertices are even; (c) there exists a regular 2-coloring of the faces compatible with the exterior domain (this means that the part of the plane complimentary to the polygon is considered white, and the faces of the polygon adjacent to its boundary are considered black).

13. (a) The mapping f is constant, that is, it takes the same value y_0 at all points of X; (b) f can be an arbitrary mapping because, for the ε-neighborhood of y_0, one can take the entire set Y.

14. Given: a disc of radius R with the center at O, a point x inside the circle lying at the distance r $(0 < r < R)$ from the center, and the point y, the projection of x on the boundary circle C (see Figure 53). Take any $\varepsilon > 0$. Let A be a common point of the circle C and the circle of radius ε with the center at y, B the projection of x on the line OA. One has $yA = \varepsilon$, $xB = \delta$. We have to find δ from given r, R, and ε. Denote by φ the value of the angle OxB, and by OM the perpendicular dropped from O on the line yA. From the triangle OxB one has $\delta = r \sin \varphi$, and from the triangle OAM it follows that $\sin \varphi/2 = \varepsilon/2R$. Hence

$$\delta = 2r \sin \frac{\varphi}{2} \cos \frac{\varphi}{2} = 2r \sin \frac{\varphi}{2} \sqrt{1 - \sin^2 \frac{\varphi}{2}} = \varepsilon \frac{r}{2R^2} \sqrt{4R^2 - \varepsilon^2}.$$

This implies, in particular, that if $r = 0$ then $\delta = 0$. Therefore, for $r = 0$ no positive ε has a corresponding δ, which again indicates that the mapping is discontinuous at the center of the circle.

17. Let us prove that the function is bounded from above, that is, prove the existence of the number M. If no such M exists, one can find a point x_1 in the closed interval $[a, b]$ such that $f(x_1) > 1$, then a point x_2 such that $f(x_2) > 2$, ..., the point x_n such that $f(x_n) > n$, etc. Since a closed interval is compact, a subsequence of points can be picked from $\{x_n\}$ converging to some point x_0 in $[a, b]$. (For convenience, we assume that

the sequence $\{x_n\}$ itself converges to x_0.) The function f is continuous at the point x_0. That is, for each $\varepsilon > 0$ there is a neighborhood of x_0 such that images of all points x of this neighborhood lie in the ε-neighborhood of the point $f(x_0)$. The latter implies that the following inequalities hold:

$$f(x_0) - \varepsilon < f(x) < f(x_0) + \varepsilon$$

(of which we need the second one). Since the sequence $\{x_0\}$ converges to x_0, all its points starting with some index n_0 belong to the chosen neighborhood of x_0, so that one has $f(x_n) < f(x_0) + \varepsilon$ for all $n \geq n_0$, thus contradicting our assumption that $f(x_n) > n$. This proof applies to any compact set, not just to a closed interval.

18. Let f be a continuous mapping of the closed interval $[a, b]$ into itself. If at least one of its endpoints is a fixed point, there is nothing to prove. Otherwise, consider the continuous function $g(x) = f(x) - x$ on $[a, b]$. We have $g(a) > 0$, $g(b) < 0$. By the intermediate value theorem there is a point x_0 such that $g(x_0) = 0$. Hence $f(x_0) = x_0$.

19. Let f be a continuous function on the closed interval $[a, b]$, and let $f(a) < c < f(b)$. Divide the closed interval $[a, b]$ in half by the midpoint $d = (a+b)/2$. If $f(d) = c$, there is nothing to prove. If, for example, $f(d) < c$, consider the closed interval $[d, b]$ and apply the argument similar to the one used in the proof of the fixed point theorem.

20. If $f(a) = g(a)$ or $f(b) = g(b)$, then everything is proved. Otherwise one has $f(a) - g(a) < 0$ and $f(b) - g(b) > 0$, and the statement follows from the fixed point theorem applied to the function $h(x) = f(x) - g(x)$ on $[a, b]$. The statement is false if g is an arbitrary continuous function on $[a, b]$.

21. Let x_1 and x_2 be such points of the closed interval $[a_1, a_2]$ that $f(x_1) = b_1$, $f(x_2) = b_2$ and suppose, to be definite, that $b_1 < a_1 \leq x_1 < x_2 \leq a_2 < b_2$. The statement follows from the fixed point theorem applied to the function $f(x) - x$ defined on the closed interval $[x_1, x_2]$.

22. Any fixed point of the mapping f is also a fixed point for $g = f^2$. We can, therefore, assume that f has only one fixed point x_0 and that $a < x_0 < b$, $f(x) > x$ for $a \leq x < x_0$ and $f(x) < x$ for $x_0 < x \leq b$. Because f maps the closed interval into itself, there is a point x_1 $(a \leq x_1 < x_0)$ such that $f(x_1) = b$, and the point x_2 $(x_0 < x_2 \leq b)$ such that $f(x_2) = a$. By the intermediate values theorem there is a point x_3 $(x_1 \leq x_3 < x_0)$ such that $f(x_3) = x_2$. Hence $g(x_3) = f(x_3) = f(x_2) = a$. If $g(a) = a$, then the mapping g has two fixed points: a and x_0. Otherwise one has $g(a) > a$, $g(x_3) = a < x_3$ and, by applying the intermediate value theorem to the function $g(x) - x$ on the closed interval $[a, x_3]$, we see that g has a fixed point x_4 $(a < x_4 < x_3)$.

If f maps the closed interval $[a, b]$ into itself, but not necessarily onto itself, then the mapping $g = f^2$ may have only one fixed point.

24. For $x \neq 0$ the equation can be written in the form:

$$x^3 \left(a_0 + \frac{a_1}{x} + \frac{a_2}{x^2} + \frac{a_3}{x^3} \right) = 0.$$

For $|x|$ very large the difference of the expression in the brackets with a_0 is small. Hence, if $x > 0$ and x is very large, then the left-hand side of the equation has the same sign as a_0, and if $x < 0$ and $|x|$ is very large, its sign is opposite to that of a_0. It only remains to note that any polynomial of the third order is a continuous function of x. A similar argument shows that if $a_3/a_0 < 0$, the equation has a positive root, and if $a_3/a_0 > 0$, it has a negative root.

25. The statement follows from the Borsuk-Ulam theorem.

26. The Borsuk-Ulam theorem can be applied only if the temperatures at the beginning and at the end of the hour are equal. Then there are indeed two moments of time having the stated property.

27. The statement is true if each of the polygons has an even number of sides. If the number of sides is odd for at least one of them, the statement holds only under the condition that there is a vertex of this polygon lying on the line passing through the centers of the polygons.

28. Draw an arbitrary line L on the plane and enclose the given curve K into the narrowest possible strip bounded by two lines parallel to L. The two lines are called *supporting* lines of the curve K. Now enclose K into the strip formed by the two supporting lines perpendicular to L. This yields a rectangle circumscribed about K. A continuous rotation of the line L by the angle $\pi/2$ interchanges these pairs. Thus, if the distance between the lines of one pair was greater than that of the other pair, the rotation makes it smaller. Hence, there must be a moment when both distances are equal, and, consequently, the intersection of the strips is a square.

32. Let a be any real number. Let $M = |a - f(a)|/(1 - \alpha)$, where α is the number from inequality (9.2). We shall prove that the mapping f takes the closed interval $[a - M, a + M]$ into itself. Indeed, let x be any point of this closed interval. Then

$$|f(x) - f(a)| \leq \alpha|x - a| \leq \alpha M,$$

and therefore

$$|f(x) - a| = |f(x) - f(a) + f(a) - a|$$
$$\leq |f(x) - f(a)| + |f(a) - a| \leq \alpha M + |f(a) - a|.$$

By the definition of the number M, one can replace $|f(a) - a|$ with $(1 - \alpha)$. Then

$$|f(x) - a| \leq \alpha M + (1 - \alpha)M = M$$

as asserted.

33. The answer is affirmative. Let us show that the given mapping f is a contraction on the entire real line. For all real numbers x and y one has

$$(S.1) \qquad |f(x) - f(y)| = \left| \frac{1}{4 + x^2} - \frac{1}{4 + y^2} \right| = \frac{|x^2 - y^2|}{(4 + x^2)(4 + y^2)}$$
$$= \frac{|x + y|}{(4 + x^2)(4 + y^2)}|x - y|.$$

Applying the well-known inequality $\sqrt{ab} \leq (a+b)/2$ (where $a \geq 0$ and $b \geq 0$) to $a = 4$ and $b = x^2$, one has

$$|x| = \frac{1}{2}\sqrt{4x^2} \leq \frac{4+x^2}{4}.$$

Hence

$$|x+y| \leq |x|+|y| \leq \frac{4+x^2+4+y^2}{4} \leq 2 + \frac{x^2+y^2}{2} + \frac{x^2 y^2}{8} = \frac{1}{8}(4+x^2)(4+y^2).$$

Substituting this estimate into (S.1), one obtains

$$|f(x) - f(y)| \leq \frac{1}{8}|x - y|$$

as required.

34. (a) No; (b) no (a continuous mapping of figure eight into itself having no fixed point can be obtained, for example, as follows: the first circle is mapped into the common point of two circles and then the second circle is rotated by a small nonzero angle).

35. (a) Yes (a disc can be projected onto a triode along the dotted lines (see Figure 54) continuously); (b) yes; (c) no, since the figure eight does not possess the fixed point property.

36. (a) Yes; (b) yes; (c) yes.

37. No.

38. No.

39. Yes. Here is an outline of the proof. Let X be a compact, Y its subset, and $f : X \to Y$ a retraction. Let $\{y_n\}$ be a sequence of points of Y. A subsequence $\{y_{n_k}\}$ can be picked from $\{y_n\}$ converging to a point x_0 of X. Each point y_{n_k} is a fixed point for f, hence x_0 is also a fixed point. Therefore, x_0 belongs to Y, so that Y is a compact.

40. Yes. For example, one can first project the square continuously on its side and then stretch this side to cover the whole boundary.

41. Assuming the contrary, let q be a point in the disc K such that $f(p) \neq q$ for all p in the disc K. That is, q does not belong to the image of the disc. Since f is identical on the circle C, the point q must lie inside K and not on its boundary C. Then the composition of the mappings f and φ is a retraction of K onto C. Because such a retraction is impossible, the statement is proved.

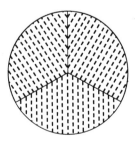

FIGURE 54

42. Assuming the contrary, let $f(p) \neq g(p)$ for all p in the disc K. Draw the ray from the point $g(p)$ to the point $f(p)$ and extend it until it intersects the boundary circle C at the point $\varphi(p)$. Now denote $f(p) = q$ and $\varphi(p) = \psi(q)$. The continuous mapping ψ sending the point q into the point $\psi(q)$, is (by the statement of the Problem 41) defined on the entire disc K, and maps it into the circle C in such a way that C is mapped into itself identically. Hence $\psi : K \to C$ is a retraction, a contradiction. The statement is false if g is an arbitrary continuous mapping of the disc into the plane.

43. A homotopy of f_0 to f_1 can be defined by the formula

$$f_t(x) = (1 - t)f_0(x) + tf_1(x), \quad 0 \leq t \leq 1.$$

44. The degree of the mapping f is 12. It equals the ratio of the velocities of the two hands. The mapping f has eleven fixed points: 0 hours, $1\frac{1}{11}$ hours, $2\frac{2}{11}$ hours, $3\frac{3}{11}$ hours, etc. The mapping φ is not defined as each position of the minute hand may correspond to different positions of the hour hand.

45. The degree of the first mapping is 2; it has one fixed point: 0 hours. The degree of the second mapping is -1; it has two fixed points: 0 hours and 6 hours. The degree of the third mapping is 1; it has no fixed points.

51. If $n = 0$, it is a mapping sending the entire sphere into one of its points, or, in general, a mapping whose image does not cover the sphere. If $n \neq 0$, consider the sphere as the surface of the Earth with the usual coordinate network on the globe. The desired mapping sends any point on the globe into the point having the same latitude and n-fold longitude.

54. There is a point x on the sphere such that $f(f(x)) = x$ (Problem 53). Then, denoting $f(x) = y$, one has $f(y) = x$. If $x = y$, this point is a fixed one; otherwise the mapping exchanges the points x and y.

References

1. M. B. Balk and V. G. Boltyansky, *Geometry of masses*, ("Kvant" series, vol. 67), "Nauka", Moscow, 1987. (Russian)
2. V. G. Boltyansky, *Method of iterations*, Kvant **3** (1983), 16–21, 37. (Russian)
3. V. G. Boltyansky and V. A. Yefremovich, *Topology through pictures*, ("Kvant" series, vol. 21), "Nauka", Moscow, 1982; German transl., *Anschanliche Kombinatorische Topologie*, VEB Deutschen Verlag Wiss., Berlin, and F. Vieweg & Sohn, Braunschweig, 1986.
4. N. Ya. Vilenkin, *Method of successive appproximations*, (Popular lectures in mathematics, vol. 35), "Nauka", Moscow, 1968; English transl. of 1st ed., *Successive approximation*, Pergamon Press, Oxford, and Macmillan, New York, 1964.
5. N. E. Steenrod and W. G. Chinn, *First concepts of topology*, Random House, New York and Toronto, 1966.
6. Hugo Steinhaus, *Kalejdoskop matematyczny*, 2nd ed., Państwowe Zakłady Wydawnictw Szkolnych, Warsaw, 1954; Russian transl., "Nauka", Moscow, 1981.
7. ____, *Problems and arguments*, "Mir", Moscow, 1974. (Russian)